Lettuce
Production Technology

莴苣生产技术

张　更　冯英娜◎主编

中国财富出版社有限公司

图书在版编目（CIP）数据

莴苣生产技术 = Lettuce Production Technology：英文 / 张更，冯英娜主编 . —北京：中国财富出版社有限公司，2023.12

ISBN 978-7-5047-8034-8

Ⅰ . ①莴…　Ⅱ . ①张… ②冯…　Ⅲ . ①莴苣—蔬菜园艺—英文　Ⅳ . ① S636.2

中国国家版本馆 CIP 数据核字（2023）第 238069 号

策划编辑	刘静雯	**责任编辑**	刘静雯	**版权编辑**	李　洋
责任印制	尚立业	**责任校对**	张营营　庞冰心	**责任发行**	敬　东

出版发行	中国财富出版社有限公司		
社　　址	北京市丰台区南四环西路188号5区20楼	**邮政编码**	100070
电　　话	010-52227588 转 2098（发行部）		010-52227588 转 321（总编室）
	010-52227566（24小时读者服务）		010-52227588 转 305（质检部）
网　　址	http: //www. cfpress. com. cn	**排　　版**	宝蕾元
经　　销	新华书店	**印　　刷**	北京九州迅驰传媒文化有限公司
书　　号	ISBN 978-7-5047-8034-8/ S · 0057		
开　　本	710mm × 1000mm　1 /16	**版　　次**	2024年9月第 1 版
印　　张	15.5	**印　　次**	2024年9月第 1 次印刷
字　　数	270千字	**定　　价**	49.80 元

版权所有 · 侵权必究 · 印装差错 · 负责调换

编写人员名单

主　编：张　更　冯英娜

副主编：黄思杰　王全智　陈　曦　吴　凯

参　编：闫征南　徐佳丽　张停林　巩子毓　李艳艳

Foreword

Lettuce, a widely cultivated vegetable in the north and south of China, plays an important role in vegetable production and market supply. Over the years, vegetable scientists have introduced and bred a series of superior lettuce varieties through long-term efforts, and greatly developed and innovated their culture techniques. So far, lettuce has achieved year-round production and supply.

Lettuce has a variety of types, mainly including stem lettuce and leaf lettuce. Stem lettuce is mainly cultivated in spring in areas with cold winters. In areas with warm winters, in addition to spring and autumn cultivation, it can also be cultivated earlier or later than normal cultivation time. In recent years, lettuce can be sown in rows and supplied all year round by using winter cold-proof and heat preservation techniques and summer sun-proof and rain-proof culture techniques in protected areas. This not only enriches the varieties of lettuce on the market but also increases economic benefits. Leaf lettuce is one of the main vegetables supplied in the spring off-season market in the Yangtze River basin from March to May. Different varieties are used for sowing and harvesting by stages in warm areas, thus fully realizing annual production and supply.

With the development of China's vegetable industry, lettuce culture techniques have changed rapidly, developing from traditional open cultivation to protected cultivation. This book mainly introduces the most common cultivation modes, including solar greenhouse lettuce cultivation, delayed autumn cultivation of polytunnel lettuce, early spring cultivation of polytunnel lettuce, NFT, deep flow technique, floating-board capillary hydroponics, aeroponics, solid substrate culture, etc. Various cultivation modes help the annual production of lettuce and enrich the food varieties on people's dining tables.

This book is prepared based on the report of the 20th National Congress of the Communist Party of China, which proposes "accelerating the construction of an agricultural power and solidly promoting the revitalization of rural industries, talents, culture, ecology, and organizations", as well as the *Guidelines on Promoting the High-quality Development of Modern Vocational Education*. It can be used as a teaching book for international students majoring in horticulture, modern agricultural technology, and agriculture-related majors in higher vocational colleges, and also as a reference for technicians, managers, and scientific researchers in relevant professional fields. It is expected that this book will be beneficial to the talent training and high-quality development of the vegetable industry.

Due to the tight schedule and the ever-changing culture techniques of lettuce, it is inevitable that there are omissions and inappropriate standpoints. We appreciate your criticism and correction for improvement in later revisions.

Contents

Module 1 Knowledge of Lettuce

[Learning Objectives]

I. Knowledge

(1) Understand the origin and evolution of lettuce;

(2) Understand the botanical characteristics of lettuce and its required environmental conditions during different development periods.

II. Skills

(1) Learn to retrieve relevant information;

(2) Understand the required environmental conditions in different growth stages of lettuce.

[Preparation]

I. Required Resources

(1) A paper library and a periodicals reading room;

(2) A digital reading room and an electronic resource library;

(3) A multimedia classroom.

II. Background Knowledge

(1) Knowledge of basic methods for literature review;

(2) Knowledge of lettuce botany;

(3) Visits to demonstration bases of lettuce cultivation and production.

[Learning Tasks]

Task 1 Origin and Evolution of Lettuce

Lettuce, also called leaf lettuce, is known for its tender leaves and leafy heads that can be eaten raw. Lettuce is native to the Mediterranean coast and evolves from wild species. As early as about 4,500 years ago, lettuce was recorded in tomb paintings in the Middle Kingdom of Egypt. Bundles of stem-lettuce-like plants, with elongated heads and lanceolate blades, were painted on the tomb wall, very similar to the variety grown in Egypt today. Lettuce was then spread from Egypt to Greece, Rome, and the entire Mediterranean region. Around the Mediterranean basin, production activities were dominated by the planting of romaine lettuce, also known as cos lettuce (possibly implying that it was originally produced on Kos Island near Türkiye), which is still one of the most common varieties of lettuce. Later, lettuce planting areas expanded from the Mediterranean region to the rest of Europe. One of Schöffer's books recorded that there were four varieties of lettuce grown in northwestern Europe. In addition, one of Pieter's paintings from 1553 depicted a plant of butterhead lettuce (the painting is currently collected in the Museum Boijmans van Beuningen, Rotterdam, the Netherlands).

Lettuce was brought to the New World by Columbus. In 1494, Peter Martyr recorded that he found lettuce on Isabela Island, only two years after Columbus' first voyage. Over the 400 years since lettuce was introduced into the United States, a wide variety of leaf lettuces have been grown there, including looseleaf lettuce with crinkled leaves and soft blades, also known as Batavia lettuce. At the beginning of the 20th century, butterhead lettuce was the most popular variety. Then, Batavia lettuce began to dominate because it could be grown on large irrigation farms in the western United States and could maintain good quality within 10–12 days for shipment to the rest of the country. In the 1940s, T.W. Whitaker hybridized iceberg lettuce (Fig. 1-1), a compact hard-head crisp lettuce. Soon, iceberg lettuce began to become popular in the United States and dominated the US market. Iceberg lettuce was later introduced into Europe and became the most important variety in the UK and Scandinavian countries. Iceberg lettuce is also popular in Spain, Germany, Australia, and Japan, where it has become a staple. Lettuce was introduced to China between 600 AD and 900 AD when the Chinese chose to grow varieties with thick, juicy, non-bitter stems and long, narrow leaves, mainly for cooking. In the Song Dynasty, Tao Gu recorded lettuce in his *Records of Pure Marvels*: In the Sui Dynasty, an envoy came to China from Guo Country (in the Western regions) with a kind of vegetable seeds. People bought the seeds at such a high price that the vegetable was called Qianjin Vegetable (Qianjin means valuable), now called lettuce. In the Song Dynasty, Su Shi recorded purple lettuce in his *Simple Discourses* on the *Investigation of Affairs and Things*. The earliest record of asparagus lettuce cultivation in China can be found in the *Essentials of Agriculture and Sericulture* edited by the Agricultural Department in the Yuan Dynasty.

Fig. 1-1 Iceberg Lettuce

Lettuce is a descendant of wild species, with no spines on its leaves and stems. It is characterized by reduced bitterness, fewer buds, slower bolting (except the stem lettuce), and larger seeds. After further artificial selection and breeding, lettuce has been changed in the size, shape, color, texture, and taste of blades and plants, heading, disease and pest resistance, and adaptability to different geographic regions and environments.

Task 2 Botanical, Flowering, and Pollination Characteristics of Lettuce

Lettuce (*Lactuca sativa* L.) is an annual or biennial herbaceous plant in the genus *Lactuca*, family Asteraceae, native to the Mediterranean Coast and domesticated from wild species. The ancient Greeks and Romans were the first to eat lettuce, which was then introduced to China via western Asia around the beginning of the 7th century. With a long planting history in China, lettuce is widely cultivated in the southeast coastal area, especially in metropolitan suburbs and Guangdong and Guangxi provinces. With the rapid expansion of planting areas in recent years, lettuce that was once supplied to hotels and restaurants can now be available to ordinary families. Lettuce is divided into leaf lettuce and stem lettuce. Leaf lettuce is also known as *Lactuca sativa* L., with blades or leafy heads as the main edible parts, while stem lettuce is also known as asparagus lettuce, with thick young stems as the main edible parts. The roots, stem, leaves, flowers, and seeds of lettuce are shown in Fig. 1-2.

Fig. 1-2 Roots, Stem, Leaves, Flowers, and Seeds of Lettuce

I. Root

Lettuce has a taproot system that forms a network of shallow roots with strong regeneration capacity, mainly distributed in the surface soil layer of 20–30 cm in depth. Lettuce cultivated by direct seeding has deeper main roots and fewer lateral roots than those cultivated by seedling transplantation.

II. Stem

In the vegetative phase, lettuce develops a dwarf stem that slowly elongates and thickens as the plant thrives. After rosette leaves appear, stem lettuce develops an enlarged succulent young stem, which consists of two parts: the embryonic axis and the flower stalk. The stem may be green, greenish-white, or purple, and in a long rod, long conical, or short rod shape. For leaf lettuce, the stem is less developed as it is short and discal.

III. Leaf

Lettuce develops loose or compact inner leaves and radical leaves that alternate on dwarf stems, with spreading or shrunken surfaces and undulate or lobed margins. The leaves may be green, yellowish-green, or purplish-red, and in an oblanceolate, oblong, or oblong-oblanceolate shape.

IV. Flower

Lettuce has a conical capitulum with about 20 ligulate florets that have pale yellow corollas and unilocular ovaries. It is self-pollinated or cross-pollinated by insects.

V. Seed

Lettuce seeds are contained in achenes. The achenes are oblanceolate, small, elongated, and flat, with silver or dark brown surfaces and 6–7 fine veins on each surface. They have an acute and mucronulate apex, with an umbrella-like slightly rough pappus that can be scattered with the wind for propagation. Lettuce seeds have a 1,000-seed weight of 0.8–1.5 g, a life span of 5 years, and a useful life of 2–3 years.

VI. Flowering and Pollination Characteristics

Lettuce is self-pollinated, with 12–25 flowers per capitulum. All simple flowers are yellowish-white ligulate complete flowers, with 5 stamens in barrel shape and

1 pistil. The simple flowers of lettuce have a very short flowering period, only 1–2 hours, and they will be fertilized just 6 hours after pollination. When anthers open to disperse pollens before flowering, ovaries elongate toward the anther tubes, making pollens attached to the tomenta on the ovaries, thus completing self-pollination. However, since the natural crossing rate can be high in dry climates, isolation in seed collection is necessary.

The corollas of lettuce simple flowers, along with stamens, styles, and stigmas, fall off 2–3 days after flowering. It takes 15–20 days from bud emergence to flowering and 13–20 days from flowering to seed maturation, which is less than the number of days required by general vegetable crops. Lettuce requires a higher temperature (22–28°C) to flower and set seeds. The higher the temperature, the shorter the time from flowering to seed maturation. If the temperature is lower than 15°C, the lettuce will bloom but cannot set seeds.

Task 3 Growth Cycle of Lettuce

The growth cycle of both leaf lettuce and stem lettuce involves a vegetative phase and a reproductive phase.

I. Vegetative Phase

The vegetative phase starts at sowing, continues through the entire process of blade development, and ends at the beginning of inflorescence differentiation. It consists of the germination stage, seedling stage, rosette stage, and organogenesis stage. The plant grows slowly until it produces two true leaves, after which the number, weight, and area of leaves begin to increase rapidly, and the plant enters its peak growth period. The duration of each stage varies with lettuce varieties and cultivation seasons.

1. Germination Stage

The germination stage lasts 8–10 days from sowing to the emergence of the first true leaf, which is also considered a critical feature of germination. The seeds germinate at a minimum temperature of 4°C and an optimal temperature of 15–20°C. The germination is uneven below 15°C. The germination rate will decrease significantly when the testa absorbs water at a temperature above 25°C, and the germination will be stunted above 30°C. Some varieties germinate fast in light, and lights vary in effects: red light promotes germination, while near-infrared light and blue light inhibit germination.

2. Seedling Stage

The seedling stage lasts 20–25 days, from the emergence of the first true leaf to the full unfolding of the first blade ring, which is also considered a critical morphological feature of the seedling, with 5–8 blades on each ring.

3. Tillering Stage

The tillering stage, also known as the rosette or rosetting stage, lasts 15–30 days from the full unfolding of the first blade ring to the full unfolding of the second blade ring. The inner leaves of head lettuce begin to compact, while looseleaf lettuce has no such stage. The rosette stage of stem lettuce can be further divided into the early stage (leaves growing slowly) and the late stage (leaves growing fast, with the number of blades increasing rapidly).

4. Organogenesis Stage

The organogenesis stage lasts 15–30 days. Head lettuce matures when inner leaves form a compact leafy head, while looseleaf lettuce is considered mature when the leaves reach a consistent height. As for stem lettuce, the organogenesis stage is also called the succulent stem formation stage. From the seedling stage to the early rosette stage, lettuce develops a dwarf stem. In the late rosette stage, the dwarf stem grows faster and expands gradually with the rapid increase in the number and area of leaves. Furthermore, the succulent stem elongates and thickens rapidly in this stage.

II. Reproductive Phase

Flower bud differentiation is a sign of lettuce transitioning from vegetative growth to reproductive growth. Lettuce has less stringent requirements for low temperature and long daylight during vernalization. It is closely related to the accumulated temperature, while the low temperature is unnecessary. Even under continuous high-temperature exposure, lettuce can bolt as long as the accumulated temperature reaches the requirement. Long daylight can accelerate development. With long daylight, the growth of lettuce can be accelerated with increasing temperature. Lettuce is sensitive to high temperatures, but the degree of sensibility varies among varieties: Early-maturing varieties are sensitive, followed by mid-maturing varieties, while late-maturing varieties are insensitive. In addition, for different varieties or even the same variety, the flower bud differentiation can also be affected by the accumulated temperature, varying with the seeding time. After flower bud differentiation, the reproductive phase begins from bolting, continues through flowering, and ends at fructescence.

Task 4 Required Environmental Conditions for Lettuce

I. Temperature

Lettuce is a Semi-cold-resistant vegetable that prefers cold and cool climates. It is slightly frost-resistant, while not heat-resistant, and will be stunted in hot seasons. With varied temperature requirements in different development periods, lettuce seeds can germinate above 4°C, and germinate faster under higher temperatures in the range of 5–28°C. At the optimal germination temperature of 15–20°C, the seeds will sprout in about 4–5 days. The germination uniformity is poor below 15°C, and the germination will be stunted below 4°C or above 30°C. Therefore, in order to improve the activity of enzymes and accelerate the transformation of other substances in seeds, the seeds to be sowed in summer can be soaked in liquid at 5–18°C for 3–5 days in advance. Lettuce is highly adaptable to temperature in the seedling stage, i.e. it not only can tolerate a low temperature of -1– -2°C but also can grow slowly above 29°C. In this stage, its optimal growth temperature is 12–20°C, 20°C during the day and 16°C at night. In the rosette stage and organogenesis stage, the optimal temperature is 11–18°C. At above 24°C, especially above 19°C at night, early bolting occurs easily to reduce product quality and economic benefits. During actual production, the diurnal temperature difference is often increased to reduce leaf respiration rate, reduce consumption, increase accumulation, and promote growth to obtain a higher yield.

In the reproductive phase, lettuce requires a higher temperature. The optimal temperature for bolting and seed setting is 22–29°C. The higher the temperature in this range, the fewer the days needed from flowering to seed maturation. At 19–22°C, the lettuce will have mature seeds 10–15 days after flowering, while at 10–15°C, it will bloom but cannot set seeds. Lettuce cultivated under cold conditions has a higher yield and better quality than those cultivated under high temperatures and drought conditions. Head lettuce has a narrow range of optimal growing temperatures and is less resistant to cold and heat than other varieties. In the heading stage, the optimal temperature for the growth of head lettuce is 17–18°C. Above 21°C, the leafy heads for sale cannot be well-formed, or inner leaves will be necrotic and rotten because of the high temperature inside. In the scorching sun, blade tips will become withered and yellow, and easily produce a bitter taste. Leaf lettuce is less resistant to cold and heat than stem lettuce, with a lower overwinter or oversummer survival rate.

II. Light

Lettuce seeds are positively photoblastic. Appropriate scattered light can hasten seed germination. For example, seeds under red light germinate faster. After being sown, the seeds exposed or covered with thin soil layers can germinate faster under suitable temperature, water, and oxygen conditions. Lettuce is a long-day plant that grows well with thick leaves under adequate light exposure. Head lettuce is more resistant to and can take full advantage of low light. Under adequate light exposure, the plant grows well with thick leaves and a compact leafy head. However, under too weak light, the assimilation of plants weakens, and the plant will develop thin lcaves and a loose leafy head, with decreased yield and quality. Therefore, dense planting should be avoided to prevent the growth of outer leaves only (without leafy heads), which affects the yield. However, head lettuce is also not adaptable to strong light. For example, direct light on the inner leaves can affect the head formation and thus reduce quality. Lettuce is insensitive to low temperatures during flower bud differentiation. Since it is a long-day plant, it will quickly bolt and flower at a high temperature and under long daylight conditions for 12–14 hours. Therefore, in order to achieve a good yield, appropriate shading and other corresponding management measures should be considered in addition to the selection of suitable varieties. Generally, early-maturing varieties are the most sensitive to photoperiod, followed by mid-maturing varieties, while late-maturing varieties are insensitive.

III. Water Content

Lettuce develops shallow roots, tender stems, and leaves with high water content, large area, and strong transpiration. It is sensitive to surface soil water and intolerant to drought. Throughout the growth period, it requires a steady and sufficient water supply. However, there are varied requirements for water supply in different development periods. In the seedling stage, the soil should be kept moist to avoid seedling ageing or spindling. In the rosette stage, the water supply should be properly controlled to promote root development so that the rosette leaves can be fully differentiated, developed, and enriched. In the organogenesis stage, the water supply should be sufficient, especially in the early stage of head formation of head lettuce, as the lack of water in this stage will lead to small, loose, and bitter leafy heads, and reduce quality and yield. In the late stage of head formation, a steady water supply is required to avoid splitting leafy heads, loss of commercial value, or occurrence of soft rot and *Sclerotinia* disease.

IV. Soil and Nutrition

Lettuce has a taproot system that forms a network of shallow roots, with weak absorption capacity and high oxygen demand. It may develop stunted roots in heavy and barren soils, and may be prone to downy mildew under rainy and waterlogged conditions. Therefore, it should be planted in loam or sandy loam with abundant organic matter, good water and nutrient holding capacity, and good permeability, preferably on high ground where it is convenient for drainage and irrigation.

Neutral or slightly acidic soil with a pH value of 6.5–7.0 is most suitable because plant growth will be affected if the pH value is above 7.0 or below 5.0.

Lettuce has a high requirement for soil nutrition since it is a fast-growing leaf vegetable with a short development period. Thus, an adequate supply of nitrogen fertilizer is necessary throughout the development period. Otherwise, blade differentiation and expansion will be inhibited, as will the growth of leafy heads. Phosphate fertilizer has an important effect on root and blade development. If phosphorus is lacking at the seedling stage, the roots will probably be stunted, with dark green leaves and a reduced number of blades. Although potassium has little effect on blade differentiation, it has a significant effect on leaf weight, as potassium can speed up the transport and accumulation of photosynthates in the leafy head. The lack of phosphorus in the heading stage may lead to loose leafy heads and a significant decrease in yield. Therefore, potassium fertilizer should be supplemented in a timely manner at the beginning of leafy head formation. Head lettuce requires nitrogen, phosphorus, and potassium in a ratio of 1 : 0.47 : 1.76 throughout its life. In addition, trace elements are also very important for lettuce and should be timely supplemented if they are insufficient in the soil. Otherwise, calcium-deficient lettuce often suffers from tipburn and thus develops a rotten leaf head (Fig. 1-3); magnesium-deficient lettuce often suffers from leaf chlorosis; boron-deficient lettuce grows slowly, with upward curved and deformed top leaves and increased spots or even plaques on blades; copper-deficient lettuce suffers chlorosis at blade edges with pink leaf veins, spreading from old to young leaves, and cannot grow a leaf head. During the growth cycle, compound foliar fertilizer may be sprayed two to three times, which can not only prevent plant diseases but also supplement trace elements needed for plant growth and development, thus avoiding the symptoms caused by trace element deficiency.

Fig. 1-3 Manifestations of Tipburn in Lettuce

[Summary]

I. Key Points

Understand the growth cycle of lettuce, and be able to regulate and control the growth of lettuce in fields depending on different growth cycles.

II. Difficult Points

Regulating and controlling temperature, light, moisture, fertilizer, and other environmental conditions during lettuce planting to produce lettuce of better quality.

[Skill Training]

Skill Training 1-1: Investigation of Lettuce Planting Worldwide

I. Purposes and Requirements

Understand lettuce planting worldwide by consulting lettuce information online, watching video recordings of visits to plantations, and exploring problems and finding solutions.

II. Planning

1. Investigation Methods

(1) Pool relevant library information and online information, and summarize.

(2) Use video recordings and equipment to present lettuce planting in different places with slides based on the lettuce cultivation videos purchased.

2. Investigation Plans

Investigate the main varieties, yield, and cropping pattern of lettuce in a country. Investigate the main ways to eat lettuce in this country.

III. Implementation

(1) Work in teams. Each team develops an investigation plan for a country.

(2) Design investigation forms, conduct field investigations, and keep records.

(3) Write an investigation report and analyze the characteristics of lettuce planting in various countries.

(4) Prepare PPT slides based on the investigation report and make a presentation in class.

[Extension Tasks]

I. Review Exercises

(1) What are the main planting areas of lettuce worldwide?

(2) Which planting area mainly grows head lettuce?

II. Case Sharing

Astronauts Grow Lettuce in the Tiangong Space Station

On November 11, 2016, the 24th day of the combination of Shenzhou-11 Spacecraft, many netizens expressed their curiosity about the scientific experiments of planting lettuce in the Tiangong Space Station. Astronaut Jing Haipeng introduced the experiments as a special correspondent of the Xinhua News Agency. Meanwhile, Wang Longji, an associate researcher from the Environmental Control and Life Support Laboratory, China Astronaut Research and Training Center, explained why lettuce was chosen for planting: First, lettuce has a growth cycle of one month, exactly the same length of time as astronauts' stay (30 days) this time; Second, we have mastered relatively mature technologies for lettuce planting on the ground; Third, lettuce is edible and can be used as a food ingredient in subsequent in-orbit experiments; Fourth, it is easier for us to disseminate scientific knowledge to the general public as lettuce is a common plant.

Astronaut's diary: We did some routine care work on the first day, including measuring the moisture content, nutrient content, and the lighting of the substrate, and injecting air into the substrate with a syringe. We have an instrument for measuring the moisture content. If the measured value is low, it means that the lettuce needs to be watered. We injected air to allow the roots of lettuce to take a fresh breath, which is conducive to plant growth. We are just like "farmers" in space, spending at least 10 minutes a day caring for lettuce. Additionally, the substrate used for planting lettuce in space differs from the soil on the earth. What we use is vermiculite.

Cultivation started on the second day after we entered the module. First, we installed the cultivation facility by assembling all its parts into a white box, just like building blocks. As introduced by Wang Longji, the parts of the white box are 3D-printed with nylon, making them lightweight. The sharp contrast between white and green makes the white box visually appealing. The white box has two instruments: one for measuring the content of moisture and nutrients in the soil, and the other for measuring plant photosynthesis under closed conditions in the late stage of plant growth. Then, we watered the substrate and planted seeds. Before entering the space, some seeds had been specially pelleted and put into white cells. Because lettuce seeds are even smaller than sesame seeds, experts deliberately coated them to make them about the same size as mung beans so we can hold them directly by hand when sowing. The coating cracks after absorbing water, but in the subsequent growth stage, we found that the coating had a slight impact on the speed of seed germination.

The way we plant seeds in space is different from what we do on the earth, where we usually plant seeds first and then water them. However, since the white cells we took into space were hard, they needed to be softened by water before the seeds could be put in. We watered them before planting seeds. After planting the seeds in the facility, we covered them with a preservative film, something like mulching film to cover crops, protect the plants, and prevent water loss.

On the fifth day, we found that the seeds had germinated. At that time, Chen Dong and I were both very happy, and we shared the good news with the staff on the ground right away. We took a lot of pictures, including those with lettuce sprouts. After the seeds germinated, we removed the mulching film and turned on the lamp mounted at the top of the white box to expose the lettuce to light. The light is a combination of red, blue, and green, mainly red. As explained by Wang Longji, lettuce absorbs red light with high efficiency, and grows well under red light; Green light is used because it shines on lettuce leaves, making them visually attractive; Blue light is great for the morphological spreading of plants.

As the lettuce grew, it turned green under light. We thinned and watered it for the first time on the sixth day after sowing. On the day of thinning, Chen Dong and I found that the lettuce plants were very fresh and looked greener than those on the earth. We thinned them with tweezers by pulling out the weak plants together with their roots and keeping two seedlings in each cell. Since the seedlings are very tender, we have to be very careful when thinning so as not to damage the seedlings left. Three days later, we began the second-time thinning and watering, leaving only one seedling per cell. Watering is not actually needed every day. Experts required watering five times in total. Each time, we did it by injecting water into the roots with a syringe. In addition to sowing, thinning, and watering, we also needed to observe and take photos of lettuce every day, as well as measure the moisture content and nutrients in the substrate. By November 11, the lettuce we planted had grown well. Watching them grow day by day, we were very satisfied.

A netizen asked questions like whether the lettuce would grow in another direction in space and how well it grew. I'd like to tell this friend that the lettuce in space grows upward as it does on the earth, but looks taller than that on the earth. Wang Longji explained that although space is gravity-free, the plants grow upward because of their phototaxis and that since they are hydrophilic and chemotropic, their roots grow towards a substrate rich in moisture and nutrients. Next Tuesday, our last day of lettuce planting in orbit, we will sample the plants with leaves and roots removed, put them in a cryogenic storage unit, and bring them back to the earth.

Some netizens are very curious about whether the lettuce we planted can be eaten. I'd like to tell them that the lettuce we grew this time is just for experimenting. I believe that through research efforts, we will be able to grow all kinds of vegetables in space for eating. Furthermore, I look forward to tasting vegetables that I plant in space. This is the first time that China has cultivated vegetables artificially in space. We are not allowed to eat them. We are going to bring back plant samples for biosafety tests to see if the microbial content on the surface of the plants exceeds the limit. Only after the test results are satisfactory will we consider eating the vegetables cultivated in the next experiment. In-orbit plant cultivation will be an essential technique for long-term manned space activities and deep space exploration in the future. We will also conduct experiments on the large-scale cultivation of other species in the future. Through several rounds of experiments, we will gradually master the rule of growing plants in space to support the future cultivation of more plant species in larger areas in the space station.

Module 2 Types and Main Varieties of Lettuce

[Learning Objectives]

I. Knowledge

(1) Understand the main types of lettuce;

(2) Understand the main varieties of lettuce grown in various regions.

II. Skills

(1) Be able to consult relevant materials by searching books and electronic databases, among others;

(2) Understand the dietotherapy effects of lettuce.

[Preparation]

I. Required Resources

(1) A paper library and a periodicals reading room;

(2) A digital reading room and an electronic resource library;

(3) A multimedia classroom.

II. Background Knowledge

(1) Understand knowledge of basic methods for literature review;

(2) Understand the main lettuce varieties in China;

(3) Observe the cultivation mode of lettuce in plant factories.

[Learning Tasks]

Task 1 Types of Lettuce

Lactuca sativa L., also called leaf lettuce in the genus *Lactuca* family Asteraceae, is an annual or biennial herbaceous plant. As a low-calorie, high-nutrient vegetable, lettuce can be eaten raw and tastes crispy and refreshing. Native to the Mediterranean Coast, it was introduced into China long ago. It is widely cultivated on the southeast coast of China, Guangdong, and Guangxi, and is also particularly common in Taiwan. Different countries prefer different types of lettuce. Americans prefer crisphead lettuce or iceberg lettuce; French prefer crisp lettuce; Nordic and British people prefer butterhead lettuce; Chinese and Egyptians prefer main lettuce varieties, which contributes to the diversity of lettuce varieties.

As there are so many lettuce varieties, classifying them will help us understand them more systematically and intuitively. Lettuce can be classified into green-leaf, white-leaf, purple-leaf, and red-leaf lettuces by leaf color: Green-leaf lettuce contains rich cellulose; White-leaf lettuce has thin leaves and high quality, and purple-leaf and red-leaf lettuces are bright-colored and tender. Non-crisp-leaf and crisp-leaf lettuces by the texture of leaves: Non-crisp-leaf lettuce, with thin leaves, is yellowish-green, soft in texture, and resistant to squeezing during transport; Crisp-leaf lettuce, with crisp and tender green leaves, breaks easily, and is not resistant to squeezing during transport. Looseleaf lettuce, head lettuce, and romaine lettuce by the growth status of leaves: Looseleaf lettuce has ruffled and deeply-incised leaves with no heads or cylindrical heads; Romaine lettuce has upright, thin and long leaves, fully or slightly incised margins, with a cylindrical head; Head lettuce grows terminal leaves into a head. Head lettuce generally comes in three types: Butterhead lettuce with obovate, smooth, soft leaves, and slightly incised leaf margins; Crisp-leaf lettuce with oboval crisp leaves, ruffled surfaces, and incised margins is more commonly cultivated than the former; Bitter-leaf lettuce with thick oblong entire leaves, semi-heading, less cultivated but more nutritious than commonly planted varieties. In foreign countries, lettuce is generally classified into three types (leaf lettuce, head lettuce, and romaine lettuce), five types (looseleaf lettuce, crisphead lettuce, butterhead lettuce, romaine lettuce, and stem lettuce), or seven types (summer crisp lettuce and buttercrunch lettuce in addition to the former five types). The classification of lettuce varies according to different criteria and in different regions.

I. Looseleaf Lettuce

Looseleaf lettuce, also known as curly-leaf lettuce, is mostly used as salad ingredients in Europe and the United States. Common looseleaf lettuces include Glasshouse, Soft-tailed, Red Sails, Simpson Elite, Grand Rapids, Rosa Red, Rosa Green, Red Skirt, Summer Green Skirt, Purple Crown, Lollo Bionda, Red Oak Leaf, Green Oak Leaf, Red Leaf , Fangni, Austrian Greenleaf, etc.

II. Head Lettuce

Head lettuce, also known as iceberg lettuce and western lettuce, is the most popular type in the United States. Common head lettuces include Kaiser, Crsipy, General, Salinas, Great Lakes 118, Great Lakes 659–700, Great Lakes 366, Olympia, Alpen, Zaokang 99, Beishan 3, Jingyou 1, Baisheng, Wins, Imperial, Queen, Ironman, Holyfield, Packer, Kenia, Iceberg, Green Lake, etc.

III. Romaine Lettuce

Romaine lettuce has long, upright, thin leaves, with no heads or compact inner leaves, and is mainly used for salads and sandwiches. Common varieties are Dengfeng, Italian Bolt-resistant All-year-round lettuce, Japanese Romaine lettuce, Spanish Green, Lollo Bionda, Olo, Meilo, Bren, Jericho, Lamb's Lettuce, etc.

IV. Butterhead Lettuce

Butterhead lettuce (Fig. 2-1) has soft, ovoid peak-green leaves with flat surfaces, ruffled middle and lower parts, and its inner leaves compact inward. With good commerciality, it smells aromatic and tastes creamy and slightly sweet. Common varieties are Boston and Niro.

Fig. 2-1 Butterhead Lettuce

V. Stem Lettuce

Stem lettuce (Fig. 2-2) has long, thin leaves and a more bitter taste than other types. The aerial stem is edible, with whitish-green skin and crisp and tender flesh. It is bright green when young and turns whitish-green when ripe. The main edible part is the tender stem, which can be eaten raw or mixed with dressing, fried, dried, or pickled. The tender leaves are also edible. The stems and leaves contain lactucin, which tastes bitter and has analgesic effects. Thanks to strong adaptability, stem lettuce can be cultivated in spring and autumn or for overwintering. It is generally cultivated in spring and harvested in summer.

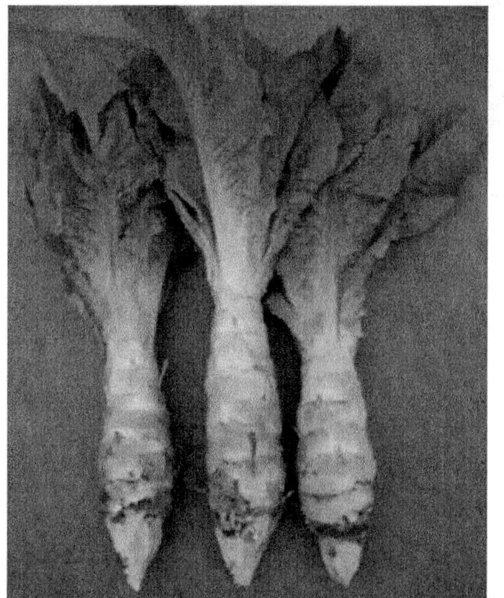

Fig. 2-2 Stem Lettuce

Task 2 Main Varieties of Lettuce

I. Looseleaf Lettuce

1. Glasshouse Lettuce

Glasshouse lettuce (Fig. 2-3) is a local looseleaf lettuce variety purified and bred by the Institute of Vegetables of Guangzhou Academy of Agricultural Sciences and introduced from Hong Kong, China. With a height of 25 cm and a spread of 30 cm, it has oval, yellowish-green, and glossy leaves that are loose and ruffled, with curly margins and a white midrib. The leaves are slightly compact inward but not tight. This type of lettuce tastes crispy and refreshing, with excellent quality. With a development period of 55 days, it is cold-resistant but not heat- or bolting-resistant, and can be cultivated in plastic tunnels or in the open field in spring and autumn, or in protected fields in winter. Considering its short development period, it can also be interplanted in protected fields. As a single plant weighs 300–500 g, it has a high net vegetable rate.

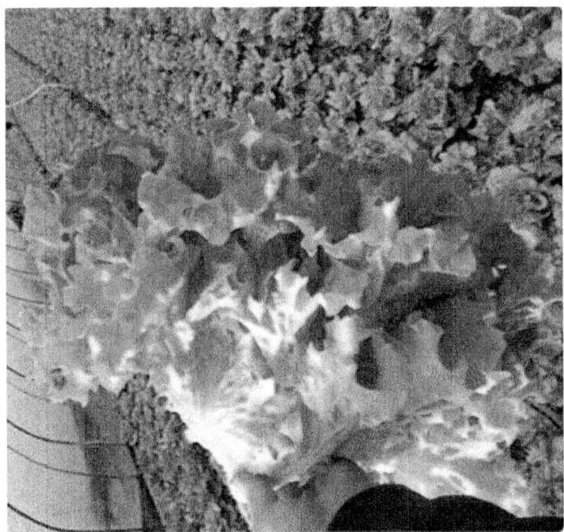

Fig. 2-3 Glasshouse Lettuce

2. Soft-tailed Lettuce

As a local variety in Guangdong, soft-tailed lettuce (Fig. 2-4) has yellowish-green and subround leaves with ruffled glossy surfaces and curly margins, slightly-inward-curved inner leaves, and white petioles. This lettuce variety is also cold- and heat-resistant with a height of 25 cm and a spread of 35 cm. An adult plant has about 28 leaves. The lettuce has thin, tender mesophyll that tastes crispy, juicy, and slightly bitter, and smells fragrant with good quality. Not resistant to high temperatures, it is planted in open fields in spring and autumn, mainly in Guangdong and Jiangxi, China.

Fig. 2-4 Soft-tailed Lettuce

3. Rosa Red Lettuce

Rosa Red lettuce (Fig. 2-5) has a very short stem and purple oblong leaves that are loose and semi-upright, with ruffled purplish-red leaf margins. Not easy to bolt, it grows 25 cm high, spreads for 20–30 cm, and looks beautiful. As a high-class variety with excellent quality, it tastes very good. Thanks to its strong adaptability, this variety is suitable for protected cultivation and open cultivation in spring, autumn, and winter.

Fig. 2-5 Rosa Red Lettuce

4. Grand Rapids Lettuce

Grand Rapids lettuce (Fig. 2-6) has yellowish-green and oval leaves that are slightly ruffled. Growing rapidly, it matures 40–45 days after being sown, about 10 days earlier than conventional varieties, and can be cultivated all year round. Crispy and tender, it has no fiber. Being resistant to heat, cold, and bolting, it grows neatly. This variety is also highly pest-resistant and adaptable and can be planted all year round in areas with a temperature above 12°C. It is suitable for open cultivation in spring and autumn and protected cultivation in winter and spring. For cultivation in summer, high-temperature prevention measures should be taken.

5. Simpson Elite Lettuce

Simpson Elite lettuce (Fig. 2-7) is a looseleaf lettuce variety that is heat-resistant

Fig. 2-6 Grand Rapids Lettuce

and bolting-resistant and can be harvested about 55 days after being sown. It has bright green leaves that are beautiful in shape. The head leaves are unfolded and curled up, with ruffled margins. It is crispy and has good quality and economic benefits. For open cultivation in spring, the seedlings should be cultivated from March to April; for open cultivation in autumn in northern China, seedlings should be cultivated from June to July; for open cultivation in autumn in southern China, seeds should be sown from July to August; for greenhouse cultivation in winter, seeds should be sown from early November to December. The planting density should be 25 cm × (20–25) cm.

Fig. 2-7 Simpson Elite Lettuce

6. Red Sails Lettuce

Red Sails lettuce (Fig. 2-8) was introduced by the Beijing Special Vegetable Seedlings Co., Ltd. from the United States in 1978. This variety is large with loose purple leaves that have ruffled surfaces and purple veins, and it looks very beautiful. As the harvest season approaches, the red color gradually darkens. Photophilous and heat-resistant, it is not easy to bolt. With a short maturity period and a development period of 50 days, it is suitable for open cultivation in spring and autumn.

Fig. 2-8 Red Sails Lettuce

II. Head Lettuce

1. Kaiser Lettuce

As an extremely early-maturing variety introduced from Japan, Kaiser lettuce (Fig. 2-9) has a development period of about 80 days. It grows compactly and neatly, suitable for dense planting in fertile soil. It has a very short center stalk in the head that is light yellowish-green and wrapped. Crispy and tender with good quality, a single head weighs about 500 g. This variety is highly resistant to heat, pest, and bolting, and forms heads well under high temperatures. Therefore, it is especially suitable for protected cultivation in spring and autumn and open cultivation in summer. The plant and row spacings are 25 cm × 45 cm, and the yield is 2,000–3,000 kg per mu (1mu=667 m^2).

Fig. 2-9 Kaiser Lettuce

2. Olympia Lettuce

As an extremely early-maturing variety introduced from Japan, Olympia lettuce (Fig. 2-10) has a development period of 65–70 days. It has light green leaves, with highly-incised margins, and a few small outer leaves. Its head is light greenish-yellow and compact, with a single weight of 400–500 g. With good quality and taste, it is highly heat-resistant and bolts very late, making it suitable for open cultivation in spring, summer, and autumn. In addition, this variety can be used as a special variety for summer cultivation. The plant and row spacings are 25 cm × 25 cm, and the yield is 3,000–4,000 kg per mu.

3. Beishan 3 Lettuce

As an early-maturing variety, Beishan 3 lettuce has a development period of 60–80 days. With a small spreading, it is suitable for dense planting (20 cm × 20 cm). As it is heat-resistant and highly pest-resistant, and bolts late, it can be cultivated all year round in Beijing except in hot July. The yield can be more than 6,667 kg per mu. It is an ideal heat-resistant variety whether cultivated in soil or water, in the open field or in a protected field.

4. Crispy Lettuce

Crispy lettuce (Fig. 2-11) is a variety originated in the United States. It has a

Fig. 2-10　Olympia Lettuce

Fig. 2-11　Crispy Lettuce

large and compact head that is yellowish-white and weighs about 800 g. The leaves are subround and relatively thick, with slightly incised margins and ruffled surfaces. The outer leaves are small in quantity and dark green in color. This variety has a high yield and good quality. It tastes crispy and sweet, and can be eaten raw and cooked. It is highly resistant to diseases such as tipburn, downy mildew, *Sclerotinia* disease, and soft rot. It can form a head only in a cold environment and is easy to bolt at high temperatures. It is better to be sown from October to December, and best to be sown from mid-October to mid-December. As an early-maturing variety, it can be harvested about 86 days after being sown. The net yield is about 4,000 kg per mu.

5. Salinas Lettuce

Originating in the United States, Salinas lettuce forms a large compact head with a few dark green outer leaves that are slightly compact inward. The head is greenish-white, nearly round, and spherical, with a single weight of about 700 g. The leaves are relatively thick with slightly ruffled surfaces. With a height of 15 cm, a spread of 40 cm, it can be planted at a proper density. With moderate disease resistance, it can form a head only in a cold environment, and it is easy to bolt at high temperatures. It is better to be sown from mid-September to December, and best to be sown in October. As an early-maturing variety, it can be harvested 83 days after being sown. With good cold tolerance, a short vegetative phase, and a large and neat head, this variety is especially suitable for protected cultivation in winter and spring.

6. Great Lakes 118 Lettuce

Originating in the United States, it is highly adaptable, with a high and stable yield and excellent quality. The head is yellowish-white and weighs about 600 g. Its outer leaves are green with slightly incised margins and slightly ruffled surfaces. With high disease resistance, it forms a head in a wide range of temperatures and bolts late. It is better to be sown from mid-September to the following mid-January. As a mid- to early-maturing variety, it can be harvested 88 days after being sown, making it suitable for open and protected cultivation in spring and autumn.

7. Great Lakes 659–700 Lettuce

Originating in the United States, it is similar to Great Lakes 118 lettuce in traits but slightly lower in yield. With a single head weighing about 500 g, it has many light green outer leaves with slightly incised margins. The maturation period varies among different plants. Moderately disease-resistant and highly cold-resistant, it grows well in a warm climate but is not heat-resistant. With a development period of 90 days, it is suitable for open and protected cultivation in spring and autumn.

8. Great Lakes 366 Lettuce

Originating in Japan, it grows vigorously, with large dark green outer leaves and a compact head that weighs more than 600 g. The harvest period should be determined according to the compactness of the head, generally two months after planting for spring cultivation, one month after planting for summer cultivation, and about three months after planting for autumn cultivation.

9. Green Lake Lettuce

Originating in the United States, it is an excellent early-maturing variety. It is less heat-resistant than Kaiser lettuce. When sown in the hot season, it grows and forms a head worse than Imperial lettuce. The head is greenish-white and oblate, with a single weight of about 0.5 kg. The outer leaves are green with slightly incised margins and slightly ruffled surfaces. With good quality and moderate disease resistance, it is better to be sown from September to the following February and can be harvested 75 days after being sown. In addition, it is suitable for open and protected cultivation in spring and autumn.

10. Alpen Lettuce

Originating in Japan, its head is greenish-white in color and oblate in shape, with a single weight of about 700 g. It has many dark green outer leaves that are subentire with slightly ruffled surfaces. This variety is 15 cm high and spreads 45 cm, with excellent quality and moderate disease resistance. It is less heat-tolerant than Imperial lettuce but more heat-tolerant than Crispy lettuce, and is easy to bolt in the hot season. It is better to be sown from mid-September to the following January. As a mid- to early-maturing variety, it can be harvested 85 days after being sown.

11. Ironman Lettuce

It is a mid-maturing variety with a development period of 82 days. It has dark green leaves with moderately incised margins. With a stable high yield, this variety forms a slightly oblate compact head that weighs about 700 g. It is resistant to rain, hot climate, tipburn, and edgeburn. It can be planted in warm seasons and is especially suitable for open cultivation in summer and protected cultivation in spring and autumn.

12. Holyfield Lettuce

As a new mid- to early-maturing variety, it has a development period of about 85 days. The leaves are bright green with slightly incised margins, and the head is round and compact with a small central stalk. It has a high yield, and resistance to heat, bolting, rainwater, tipburn, and edgeburn. This variety is suitable for cultivation in warm seasons, especially for open cultivation in spring, summer, and autumn.

13. Kenia Lettuce

As a new mid-maturing variety, it has a development period of about 85 days. Its leaves are dark green, and its head is round and uniform. With a high yield, it is both heat- and cold-resistant but not rainwater-resistant in the late stage. With good quality and a high net yield, it is suitable for slicing. It is suitable for protected cultivation or open cultivation in both northern and southern China when the weather changes from cool to hot.

14. Imperial Lettuce

As an early-maturing variety introduced from the United States, it has small and ruffled outer leaves with incised leaf margins. Weighing about 500 g, its medium-sized light-green leafy head is oblate (flat at the top) and compact, and tastes crispy and refreshing with high quality. This variety is highly heat-resistant and bolts late, making it suitable for open cultivation in spring, summer, and autumn and protected cultivation in all seasons. It features a development period of about 80 days, a spacing between plants of 30 cm × 30 cm, and a yield of 3,500–4,000 kg per mu.

15. Hydrangea-like Lettuce

It has round and light-green outer leaves with ruffled surfaces and yellowish-green inner leaves that form a compact hydrangea-like head, with a height of about 25 cm, a diameter of about 20 cm, and a weight of 600–1,000 g. With strong cold resistance and slightly weak heat resistance, it bolts late and can be cultivated throughout the year, preferably in early autumn and for overwintering cultivation. It is harvested about 50 days after planting, with a harvest period of more than 50 days. This variety is resistant to *Sclerotinia* disease, soft rot, and leaf scorch, and is currently one of the best head lettuces for sale.

16. Wins Lettuce

As a mid-maturing to early-maturing variety, it grows vigorously with a compact plant form, thick and green leaves (with slightly incised margins), and a weight of 500–600 g. Thanks to its excellent resistance to cold and heat, this variety can grow normally under both low and high temperatures, making it suitable for protected cultivation in winter and spring and open cultivation in early summer. It features a development period of 80–85 days, a spacing between plants of 30 cm × 30 cm, and a yield of 3,000 kg per mu.

17. Green-white Head Lettuce

It is also known as Beijing Iceberg lettuce, a local variety of Beijing that is classified as crisp-leaf head lettuce. With semi-erect leaves, the plant can be about 15 cm high and spread laterally for about 25 cm. Outer leaves are dark green and thick with wavy leaf margins and ruffled leaf surfaces. The subround compact leaf head has good quality and weighs about 500 g. Resistant to cold but not heat, this variety is suitable for protected cultivation with a yield of 2,000–2,500 kg per mu.

18. Wanli Iceberg Lettuce

As an early-maturing variety, it has yellowish-green out leaves that are about 21 cm long and 22 cm wide with slightly ruffled surfaces and incised margins. With a weight of 400–500 g, the subround compact leaf head is yellowish-white and tastes crisp and sweet with little fiber. This variety is resistant to heat and *Sclerotinia* disease, so it is suitable for cultivation throughout the country, with a yield of 2,000–2,300 kg per mu.

III. Romaine Lettuce

1. Romaine Lettuce

Introduced from Italy, it can grow vigorously and throughout China. Being relatively erect, the plant grows 25–30 cm high and spreads laterally for 20–25 cm with a single head weight of 250–400 g. The dark-green obovate leaves with slightly ruffled surfaces are 20–22 cm long and 15–16 cm wide, with a crisp taste and fragrant smell. This variety features a vegetative phase of 60–70 days, a spacing between plants of 25 cm × 25 cm, and a yield of 2,000–2,500 kg per mu, suitable for protected and open cultivation in spring and autumn.

2. Japanese Mibuna Lettuce

Introduced from Japan, it grows vigorously with a strong branching ability. Young leaves may grow between leaf axils, and a single plant may have 600–800 leaves. Being relatively upright, the plant may grow 40 cm high with dark-green oval leaves (no incisions on the leaf margin), elongated and greenish-white petioles, and a weight of 200–1,000 g. This variety features a vegetative phase of 60–90 days and a yield of 4,000–5,000 kg per mu.

3. Niu Li Lettuce

It is a local variety of Guangdong Province. It grows uprightly, with a plant height of 28–35 cm and a lateral spread of 30–35 cm. With a weight of 300–350 g, the plant has yellowish-green obovate leaves with slightly ruffled surfaces and wavy margins, and its inner leaves are loose with slightly poor quality. Resistant to cold but not heat, this variety features a vegetative phase of 65–80 days and a yield of about 2,000 kg per mu, suitable for cultivation in spring and autumn as well as for overwintering cultivation.

4. Dengfeng Lettuce

Mostly cultivated in Guangdong Province, it grows upright, reaching about 30 cm, and spreads laterally for about 36 cm. Classified as romaine lettuce, it weighs about 330 g and has light-green subround leaves with wavy margins and loose inner leaves.

5. Glasshouse Lettuce

As a local variety of Guangzhou, with fascicled leaves, the plant can be about 25 cm high and spread laterally for about 30 cm. It has compact inner leaves and yellowish-green, glossy, thin subround leaves that are about 18 cm long and 17 cm wide, with ruffled surfaces and wavy margins. Crisp with little fiber, it is of good quality and weighs 200–300 g. This variety features a vegetative phase of 60–80 days and a yield of 2,000–2,500 kg per mu. Resistant to cold and slightly resistant to heat, it is suitable for cultivation in spring and autumn.

6. Year-round Lettuce

The plant grows compactly, up to 18–24 cm high, and spreads laterally for 23–28 cm.

It has yellowish-green glossy subround leaves that are 17–23 cm long and 16–22 cm wide, with flat and wide petioles and yellowish-white leaf skirts (about 1 cm long). The plant may form a semi-compact head in cold seasons. Being crispy and tender with little fiber, it tastes good and weighs 200–350 g. This variety features a vegetative phase of 60–75 days and a yield of 2,000–3,000 kg per mu. With heat resistance and bolting resistance, this variety is suitable for cultivation throughout China and can be cultivated all year round in places like Guangzhou.

IV. Butterhead Lettuce

1. Boston Lettuce

Introduced from the United States, it is the leading variety of greenhouse hydroponic lettuce. As an early-maturing variety, it can be harvested 35–40 days after planting. The semi-heading plant grows vigorously with a height of about 20 cm and a lateral spread of about 30 cm. Since its green oval leaves are glossy, fleshy, and tender with high quality, this variety is commonly known as butterhead lettuce. Its outer leaves are loose, while its inner leaves are compact or non-compact, with entire leaf margins. With cold resistance and slight heat resistance, this variety adapts to extensive types of soil, with a high and stable yield and a single plant weight of about 300 g.

2. Niro Lettuce

As a mid- to early-maturing butterhead lettuce variety of Italy, it features a development period of about 60 days, a plant height of 25–30 cm, and a plant weight of 300–400 g. It has bright-green fascicled leaves that are subround, glossy, and tender, with a pleasant look. With good resistance to bolting and diseases, this variety is suitable for both hydroponic and soil cultivation.

V. Stem Lettuce

1. Pointed-leaf Lettuce (Fig. 2-12)

It is a variety selected and bred by the Chongqing Academy of Agricultural Sciences. Being a purified variety, it has green lanceolate leaves and cylindrical stems, suitable for all-season cultivation. Generally, the yield is 2,000–2,500 kg per mu.

Fig. 2-12 Pointed-leaf Lettuce

2. Erbaipi Lettuce (Fig. 2-13)

It is a variety selected and bred by the Chongqing Academy of Agricultural Sciences. As a purified variety, it is of high quality with green round leaves and greenish-white skin, with a single stem weight of 600 g. It can be harvested 110–120 days after being planted, with a yield of 2,000–2,500 kg per mu.

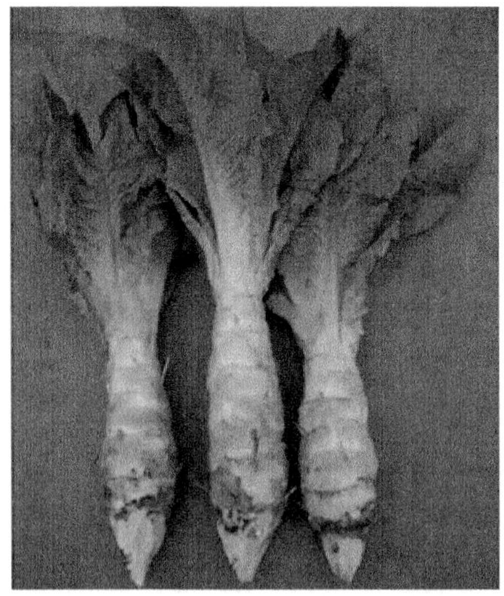

Fig. 2-13 Erbaipi Lettuce

3. Round-leaf Lettuce (Fig. 2-14)

This variety features long upright obovate leaves that are blunt and round at the apex and wrinkled on the surface. With greenish-white skin and white flesh, the stem weighs 250–300 g with good quality, a crispy taste, and a fragrant smell. With a vegetative phase of 180 days, it is mainly cultivated in spring.

Fig. 2-14 Round-leaf Lettuce

4. Red Lettuce (Fig. 2-15)

It is a traditional local variety of Sichuan Province. With good quality, its stem is green, and it weighs 600–700 g, even up to 1 kg. As an early- to mid-maturing variety, it can be harvested 100–105 days after sowing. It is highly fertilizer-tolerant, disease-resistant, and adaptable, with a yield of 2,000–2,500 kg per mu.

5. Green-skin Lettuce

As an early-maturing variety, it tends not to be hollow and is heat-resistant and slightly bolting-resistant. The green obovate leaves are 23–30 cm long, and the good-quality fleshy stem is tender and smells fresh with thin and light-greenish-white skin.

Fig. 2-15 Red Lettuce

6. Yong'an Lettuce

Affected by the specific growing conditions such as geography and climate, Yong'an lettuce is highly edible with a unique flavor and superior in quality owing to its following characteristics: bright-green thick stems with thin skin, crispy, tender, and juicy flesh with little fiber, strong aroma, low likelihood of hollowing, and low content of nitrite.

7. Universal Jade Lettuce

It has large pointed leaves, thin green skin, light-green, fragrant, and crispy flesh, with long internodes. With heat resistance and cold resistance, it is suitable for planting throughout China in all seasons. The lettuce seeds can be sowed in the Yangtze River basin from February 1st to November 30th. If sowed above 25°C from May to August, they need to be pre-germinated at low temperatures. The seedling age ranges from 25 to 30 days and the spacing between plants is 36 cm × 36 cm. High-border cultivation is recommended, and sufficient base fertilizer should be applied.

8. Yong Rong No. 1 Lettuce (Fig. 2-16)

It is a red pointed-leaf lettuce variety launched by Fuzhou Yongrong Seed Co., Ltd. It has good cold resistance and is suitable for cultivation in autumn. This variety grows rapidly, with large upright stems, with high salability, high yield, and high adaptability. A more stable yield can be achieved under suitable cultivation conditions.

Fig. 2-16　Yong Rong No. 1 Lettuce

VI. Other Lettuces

1. Huaye Lettuce

Also called endive, it grows 25 cm high and spreads laterally for 26–30 cm. It has long oval semi-erect leaves that are deeply incised at the margin and twist vertically into a comb shape. The outer leaves are green, while the inner ones are light-green to yellowish-white, gradually straight, with a light-green midrib and a white base. A single plant weighs about 500 g. With a development period of 70–80 days, this variety is of good quality, tastes slightly bitter, and is highly adaptable, heat-resistant, and less likely to be affected by diseases and pests. It is suitable for open cultivation and plastic tunnel cultivation in spring, summer, and autumn.

2. French Crisp Lettuce

It is also known as Summer Crisp or Batavian lettuce, a variety between Crisphead and Looseleaf lettuce. With thick and crispy leaves, it is relatively resistant to bolting and has a higher demand for fertilizers. A whole plant of this variety is heavier than that of other varieties. It is suitable for open cultivation or plastic tunnel cultivation in spring and autumn and protected cultivation in summer and winter.

3. Buttercrunch Lettuce

It is a hybrid of butterhead lettuce and romaine lettuce, with the characteristics of them both. However, it has fewer and more upright leaves and is less susceptible to diseases thanks to its enhanced resistance.

[Summary]

I. Key Points

Understand the types of lettuce.

II. Difficult Points

Distinguish the morphological characteristics and keys to the planting of different types of lettuce.

[Skill Training]

Skill Training 2-1: Study of Nutrition and Health Benefits of Lettuce

I. Purposes and Requirements

To understand the nutritional quality and health benefits of lettuce by consulting both domestic and foreign literature, including books and videos.

II. Planning

(1) Investigation on the nutrient content of lettuce: Investigate the type and quantity of nutrients contained in different types of lettuce.

(2) Investigation on health benefits of lettuce: What are the specific health benefits of leaf lettuce and stem lettuce?

III. Implementation

(1) Develop an investigation plan in groups for looseleaf lettuce, head lettuce, romaine lettuce, and butterhead lettuce.

(2) Design investigation forms, conduct field investigations, and keep records.

(3) Analyze the nutrient content and health benefits of different types of lettuce.

(4) Prepare relevant investigation reports. Describe the investigation method and the source of lettuce purchased.

(5) Prepare PPT slides based on the investigation report and make a presentation in class.

[Extension Tasks]

I. Review Exercises

(1) What kinds of lettuce are suitable for cultivation in different regions of China?

(2) Think about the differences in the type, content, and health benefits of nutrients in different types of lettuce.

II. Case Sharing

Health Benefits of Lettuce Dishes

1. Stir-fried Lettuce

This dish offers many health benefits and can be used for analgesia, hypnosis, diuresis, lowering cholesterol, adjuvant treatment of neurasthenia, promoting blood circulation, and anti-virus treatment.

2. Stir-fried Lettuce with Garlic

In addition to the health benefits of stir-fried lettuce, this dish also has antiseptic, anti-inflammatory, hypoglycemic, and brain-tonifying effects. Lettuce is rich in vitamins that help prevent gingival bleeding and vitamin C deficiency. Garlic has detoxifying, dredging channels and collaterals, and stomach-strengthening effects, and can serve as dietotherapy for treating epigastric crymodynia, dysentery, diarrhea, phthisis, pertussis, cold, and malaria.

3. Lettuce in Oyster Sauce

Oyster sauce is not oil but the filtered and concentrated soup left from the stewing of dried oysters. It is a kind of seasoning that contains multiple nutrients and tastes delicious. Therefore, in addition to lowering blood lipid and blood glucose, and promoting mental development and anti-aging, this dish also boosts diuresis, promotes blood circulation, helps anti-virus treatment, and prevents and treats heart diseases and liver diseases. Salt is not needed when this dish is prepared.

4. Stir-fried Lettuce with Tofu

Lettuce cooked with nutritious tofu is a high-protein, low-fat, multi-vitamin dish, with effects such as purging liver and gallbladder, tonifying yin and replenishing the kidney, whitening the skin, losing weight, and bodybuilding. It also serves as a dietotherapy in treating swelling and pain, cough with lung heat, diabetes, and splenasthenic abdominal distention, among others. Mushrooms contain rich proteins that are easy to be absorbed by the human body and can tonify the spleen, supplement qi, moisten dryness, and resolve phlegm. When cooked with lettuce, this dish is effective in treating heat cough, excess phlegm, chest tightness, vomiting, and diarrhea.

In terms of dietotherapy, stir-fried lettuce is not as good as lettuce in oyster sauce and stir-fried lettuce with mushrooms. Stir-fried lettuce with tofu is highly nutritious, but gout patients should eat with caution because tofu contains a high level of purines. Lettuce in oyster sauce and stir-fried lettuce with mushrooms are fitter for human consumption and have better dietotherapy effects than stir-fried lettuce with garlic. Patients with glaucoma, cataracts, and other eye diseases should not eat stir-fried lettuce with garlic too frequently because long-term intake of garlic will cause damage to the liver and eyes.

Module 3 Lettuce Seedling Raising Technique

[Learning Objectives]

I. Knowledge

(1) Understand the meaning of seed quality;

(2) Master the conditions of lettuce seed germination;

(3) Master the method of determining seed viability.

II. Skills

(1) Learn to retrieve relevant information;

(2) Master the method of determining lettuce seed viability;

(3) Master the method of treatment before seed pre-germination.

[Preparation]

I. Required Resources

(1) A paper library and a periodicals reading room;

(2) A digital reading room and an electronic resource library;

(3) A multimedia classroom.

II. Background Knowledge

(1) Master the basic methods for literature review;

(2) Understand the relevant knowledge about seed quality;

(3) Understand the theoretical knowledge about seed viability determination;

(4) Understand the method of treatment before seed pre-germination.

[Learning Tasks]

Task 1 Seed Quality

Seed quality means the variety quality and sowing quality of seeds in a broad sense. Variety quality is associated with the genetics (i.e. internal quality) of seeds, which can be summarized in two words: authentic and pure. Sowing quality is related to the field emergence (i.e. external quality) of seeds after sowing, which can be summarized in five words: clean, strong, full, healthy, and dry.

Authentic: It refers to the degree of authenticity and reliability of seeds, which can be expressed as authenticity. If seeds lose their authenticity, they will no longer be the original required superior seeds, and farming will be delayed.

Pure: It refers to the degree of consistency of typical characteristics of a variety, which can be expressed as variety purity. A high variety purity indicates a bumper harvest because of the excellent characteristics of this variety. In contrast, a low variety purity indicates a significantly lower yield because of the degraded quality of seeds mixed with other crop seeds or impurities.

Clean: It refers to the degree to which the seeds are clean, which can be expressed as seed purity. High seed purity indicates a low content of impurities and a high number of qualified seeds. Seed purity is one of the indicators used to calculate the sowing amount.

Strong: It refers to the rate of seed germination and seedling emergence, which can be expressed by germination ability, viability, and vigor. Seeds with a high germination rate and viability germinate neatly, and seeds with high vigor indicate a high field emergence rate and strong growth of seedlings. At the same time, the sowing amount of such seeds per unit area can be appropriately reduced. Germination rate is also one of the indicators of the sowing amount.

Full: It refers to the plumpness of seeds, which can be expressed as thousand-seed weight. A full seed indicates abundant substances inside, which is conducive to seed germination and seedling growth.

Healthy: It refers to the health condition of seeds, usually expressed as the rate of infection by diseases and pests. Seed diseases and pests directly affect seed germination rate and field emergence rate, as well as the growth and yield of crops.

Dry: It refers to the dryness of seeds, which can be expressed as seed moisture content (%). Low moisture content is conducive to the safe storage of seeds and maintenance of seed germination rate and vigor. Therefore, seed moisture content is closely related to the sowing quality of seeds.

Seed quality inspection covers seed authenticity, variety purity, seed purity, germination ability (viability), vigor, thousand-seed weight, bulk density, moisture content, and health status. In the seed quality grading standard, the major indicators are variety purity, seed purity, germination rate, and moisture content.

1. Determination of Seed Water Content

The water content of seeds is expressed as a percentage of water in the seeds on a mass basis. It is one of the indicators for safe storage, transportation, and grading of seeds. It is calculated by the following formula:

$$\text{Seed water content} = \frac{(\text{Seed mass before drying} - \text{Seed mass after drying})}{\text{Seed mass before drying}} \times 100\%$$

2. Determination of Seed Purity

Seed purity is expressed as a mass percentage of seeds of the target variety in the sample size. Seeds of other varieties or types, sediment, floral organs, and other debris are impurities. It is calculated by the following formula:

$$\text{Seed purity} = \frac{\text{Total size of test sample} - (\text{Impurity mass} + \text{Mass of seeds of other varieties or types})}{\text{Total size of test sample}} \times 100\%$$

3. Determination of Seed Fullness

The seed plumpness is normally expressed as the mass (g) of 1,000 seeds, which is called the thousand-seed weight or absolute mass of seeds. It reflects the breeding level and storage of seeds: the higher the absolute mass, the fuller the seeds, and the better the sowing effects. It also serves as a basis for estimating the sowing amount.

4. Determination of Seed Germination Ability

The seed germination ability is determined according to two indicators: germination rate and germination potential, which can be measured by the germination test. The seed germination rate is the percentage of normally germinated seeds in the total number of tested seeds under the most suitable environmental conditions for germination, reflecting the viability of the seeds. It is calculated by the following formula:

$$\text{Seed germination rate} = \frac{\text{Number of germinated seeds}}{\text{Number of tested seeds}} \times 100\%$$

Germination potential is a measure of seed germination speed and germination uniformity and is expressed as the percentage of seeds germinated within the specified days in the tested seeds. High germination potential indicates that seeds germinate quickly and neatly. The germination potential test of melons, beans, cabbages, wild cabbages, lettuces, and root vegetables lasts three to four days, and that of green onions, leeks, spinach, radishes, celeries, and eggplants lasts six to seven days. The germination potential is calculated as follows:

$$\text{Seed germination potential} = \frac{\text{Number of germinated seeds within the specified days}}{\text{Number of tested seeds}} \times 100\%$$

5. Determination of Seed Viability

Seed viability refers to the germination potential of seeds or the vitality of embryos. The key to determining the viability of seed samples, especially dormant seeds, is to estimate it quickly. For some plants with a long seed dormancy period, their seed viability must be determined rapidly to determine the seed quality within a short period. Due to the lack of equipment or time to determine the germination potential through conventional germination tests, viability must be determined to predict the seed germination potential on this basis.

Seed viability is usually expressed as the percentage of the number of viable seeds to the total number of seed samples (i.e. survival rate). It is generally evaluated according to the staining reaction of the embryo (and endosperm) after seeds are soaked in a chemical reagent solution. The main chemical test methods include tetrazole staining, indigo staining, and iodine-potassium iodide staining. Among them, the most common viability test method that has been included in the *International Rules for Seed Testing* is tetrazole staining.

Task 2 Seed Germination Conditions and Lettuce Seed Treatment

Whether a seed can germinate normally and develop into a healthy plant after germination depends on two factors: the physiological conditions of the seed itself and the external environment for seed germination. Only seeds living through the dormancy stage or non-dormant viable seeds can normally germinate and develop into seedlings under suitable conditions. Suitable environmental conditions for seed germination include adequate water supply, suitable temperature, and oxygen supply. Light, chemicals, soil, and biological factors also have certain impacts on germination.

I. Water Content

1. Minimum Water Demand for Seed Germination

Water is a prerequisite for seed germination. Seeds will not shift from an inactive state to an active state until they absorb water and will not germinate until they absorb a certain amount of water. Different kinds of seeds require different amounts of water for germination, which is expressed as different minimum water demands. The minimum water demand for germination, also expressed as water content, refers to the ratio of the minimum amount of water in the seed at the time of germination to the original weight of the seed.

2. Factors Affecting Water Absorption of Seeds

Water absorption and absorption rate of seeds are mainly affected by the chemical composition of seeds, testa permeability, external water regime, and temperature.

(1) Chemical composition: The water demand for seed germination is closely related to the chemical composition. Starch seeds and oily seeds need less water for germination, while high-protein seeds need more. In addition, seeds with a high water demand for germination generally germinate at a low rate.

(2) Testa permeability: The water permeability of testa varies greatly in seeds. For bean seeds, water mainly enters the interior through the germination port of the testa; for peas, the water permeability of umbilici is twice as high as that of other parts, while hard seeds cannot germinate due to the impermeability of the testa.

(3) External water regime: Seed water absorption is highly related to the external water regime. Some seeds can absorb enough water from saturated or nearly saturated air to germinate. Generally, seeds absorb liquid water for germination. Seeds buried in the soil can absorb water from the surrounding soil layers with a diameter of about 1 cm. Their water absorption decreases with the increase of the water absorption potential and osmotic pressure.

(4) Temperature: The temperature will obviously affect the water absorption rate of seeds at a certain stage of water absorption. The water absorption rate generally increases by 50%–80% for every 10°C rise in ambient temperature.

3. Imbibition Injury and Imbibitional Chilling Injury of Seeds

When a dry seed is just exposed to water, some substances in its cells will extravasate, such as soluble sugars, organic acids, amino acids, and inorganic ions, due to an incomplete cell membrane system, but this will be gradually alleviated with the repair of the seed cell membrane. However, for some seeds such as soybeans and French beans that have thin testa, high protein content, and strong water absorption potential, if the seed imbibition rate is too high, their cell membrane cannot be repaired and more injuries will occur such as aggravated extravasation of substances, and declined germination and growth ability of seeds, which are called imbibition injuries. Therefore, seeds, especially those with incomplete testa (such as soybeans) should not be soaked before being sowed.

For safe germination of seeds, there are certain requirements for the imbibition temperature. For dry seeds of some corps (water content within 12%–14%, depending on types of crops), if they absorb water at a low temperature above 0°C for a short time, the embryo will be damaged, making the seeds unable to germinate or develop into seedlings normally even under normal conditions, which is called imbibitional chilling injury. Imbibitional chilling injury is closely related to the serious injury of the embryo cell membrane. It occurs in the initial stage of the imbibition of dry seeds. The lower the original water content of seeds, the more vulnerable to chilling injury they will be. The threshold temperature for imbibitional chilling injury is 15°C or 10°C. Seeds of soybeans, French beans, tomatoes, eggplants, peppers, and many ornamental plants are vulnerable to imbibitional chilling injury. Before sowing at a low temperature, these seeds should be soaked in warm water (for a certain period of time, depending on the types of seeds) to help alleviate the imbibitional chilling injury.

II. Temperature

There are certain temperature requirements for seed germination, which can be summarized into the minimum, optimal, and maximum temperatures. At the minimum and maximum temperatures, at least 50% of the seeds can germinate normally; at the optimal temperature, seeds can germinate rapidly and achieve the highest germination rate. Seeds of most crops can germinate well within 15–30°C, but different crops have different temperature requirements.

The temperature requirement for seed germination is related to the growth habits of crops and the long-term ecological environment. The optimal temperature for seed germination of tropical crops is generally higher than that of temperate crops. The minimum, optimal, and maximum temperatures are 6–12°C, 30–35°C, and 40°C respectively, for thermophilous crops or summer crops, and 0–4°C, 20–25°C, and 40°C, respectively, for cold-resistant crops or winter crops.

Most seeds can germinate in a wide temperature range, but some require strict temperature conditions for germination. For example, the optimal temperature for seed germination is 15°C for celery and should not exceed 20°C for leek, lettuce, *Artemisia youngii*, and spinach. Therefore, the seeds of these winter crops are difficult to germinate in the hot summer.

III. Oxygen

Most seeds cannot germinate without oxygen. When seeds germinate, their strong aerobic respiration, in addition to some enzyme activities, requires adequate oxygen supply. The oxygen supply to the embryo during germination is affected by the external oxygen concentration, the solubility of oxygen in water, the oxygen permeability of the testa, and the oxygen affinity of enzymes.

The partial pressure of oxygen in the atmosphere is 21%, which can fully meet the needs of seed germination. Seed germination may be blocked if the seeds are buried too deep or there is more water and less oxygen in the soil, or inhibited when the oxygen concentration in the soil is lower than 9%–10%. Generally, the main factors restricting oxygen supply are water and testa. At 20°C, the diffusion rate of oxygen in the air is 10 times that in water. At the time of imbibition, the epidermal water film is thickened, making it more difficult for oxygen to diffuse to the embryo. If there is too much water during the germination of seeds of some crops, including watermelon, pumpkin, and spinach, which already have poor testa permeability, the oxygen supply will be impeded, seriously affecting germination. In this case, too much water should be avoided. Nevertheless, the oxygen supply and the oxygen demand of the embryo will change with the ongoing germination, especially after the radicle penetrates the testa, when the oxygen supply barrier disappears.

IV. Other Factors

1. Light

The impact of lighting on seed germination varies among plant species, and for most plants, its impact is minimal or ineffective. However, the photophilous seeds of certain plants, such as lettuce and celery, can germinate quickly when buried in thin layers of soil or exposed while being sowed under a proper temperature and with sufficient supply of water; other types of seeds whose germination is inhibited by light, such as cress, delphinium, onion, and amaranth, can germinate under low light or dark conditions.

2. Dormancy

Viable seeds may not germinate even under suitable environmental conditions. Probably, they are in dormancy. Seed dormancy is the result of long-term natural selection. In the temperate zone, the seeds that mature in spring will germinate immediately and grow into seedlings the same year. In contrast, the seeds that mature in autumn need to overwinter before germinating the next spring; otherwise, the seedlings will be frozen to death. The seeds of many deciduous fruit trees are subject to natural dormancy. The main causes of seed dormancy are structural disorders of the testa or skin, embryo hypoplasia, and chemical inhibition. Seed dormancy is conducive to the adaptation of plants to the external natural environment to maintain reproduction, but this brings some difficulties to sowing and seedling raising because seeds need to break dormancy before germination by after-ripening or using chemical agents in a low-temperature and humid environment.

V. Treatment of Lettuce Seeds

Lettuce seeds prefer a cool climate to a hot one. Generally, they germinate at over 4°C, germinate well at a temperature of 15–20°C, and are inhibited in germination when the temperature is above 30°C. Therefore, no treatment of lettuce seeds is required before sowing except in the high-temperature season. In July and August, pre-germination is generally required as follows: Soak the seeds in warm water at 25–30°C for 6–8 hours, then take them out, slightly dry them and perform pre-germination at 15–20°C; when 80% of the seeds are germinating, they can be sown; or first soak the seeds for 2 hours, then wrap them with wet gauze and hang them in a well or cave or place them in a refrigerator at 5–10°C for 3–4 days for low-temperature pre-germination. Then, take them out every day to wash them with clean water. When most seeds are germinating, they can be sown. In addition to the above germination methods, seeds can also be treated with plant hormones, for example, soaking in cytokinin 100 μL/L (ppm) solution for 3 minutes or in gibberellin 5 μL/L (ppm) solution for 6-7 hours, which yields good germination results.

Task 3 Lettuce Sowing and Seedling Raising

Lettuce seeds are small and have weak germinating potential, so transplanting is generally adopted in production.

I. Seedling Raising Period

Sowing and seedling raising can be performed from late July to late August for autumn cultivation, from October to November for winter greenhouse cultivation, in the middle and late February for spring cultivation in small arch sheds, and after early April for spring open cultivation. For seedling raising in the hot season, sunshade nets should be provided for shade and cooling.

II. Variety Selection

Early-maturing varieties should be selected for open cultivation in spring, heat-resistant or looseleaf (curled) varieties for cultivation in summer, and low-temperature varieties with strict requirements for temperature (cold climate during heading stage) for cultivation in autumn.

III. Seed Treatment

1. Dry Seed Sowing

The seeds can be treated with 75% chlorothalonil wettable powder weighing 0.3%–0.5% of their dry weight before being sown and then sown right after this treatment.

2. Wet Seed Sowing

Seeds are prone to hot dormancy if sown in hot summer, which inhibits their germination. Therefore, they may be soaked in the following procedures for germination in actual production: Soak the seeds in clean water at a temperature of 15–18°C for 8–12 hours, take them out and wrap them with moist towels for pre-germination at 15–20°C. The seeds will germinate in 2–3 days. Seeds can also be treated with plant growth regulators, for example, soaking in cytokinin 100 mg/L solution for three minutes, or gibberellin 1,000 mg/L solution for 2–4 hours. When 70% of the seeds are germinating, they can be sown.

IV. Preparation of Seedbed and Nutrient Soil

The seedbed quality is highly related to the strong growth of seedlings, and full preparation should be made before seedling raising. The seedbed should lie in the east-west direction in high and dry land that is leeward and sunny, with abundant humus and convenient drainage, irrigation, and transportation conditions.

Nutrient soil, also called compost, is essential for raising strong seedlings. It is required to be clean, properly structured, free of diseases and pests, with a sufficient supply of nutrients, and strong water and nutrition retention capacity that can meet the needs of seedling growth. Generally, the culture soil is prepared based on a proportion of six portions of garden soil to four portions of decomposed and sieved dung or compost for general seedbeds, and seven portions of garden soil to three portions of dung or compost for seedbeds for seedling separation and transplanting. In addition, 400–600 g of urea or ammonium sulfate, 800–1,000 g of potassium phosphate or potassium nitrate, or 1,000–1,500 g of ternary compound fertilizer should be added to each cubic meter of culture soil. The seedbed thickness should be 10–12 cm.

Organic fertilizers such as dung or compost that are used to prepare nutrient soil can be obtained in situ, but they must be fully decomposed in advance and preferably stacked from summer. Generally, an appropriate amount of water or human excreta should be sprayed for each layer of organic fertilizer stacked. A total of four to five layers should be stacked and turned every one to two months. After full fermentation in summer and autumn, the stacks of fertilizers should be mashed and sieved before they freeze for later use. For better efficiency of phosphatic fertilizer, 3–5 kg of calcium superphosphate can be added per cubic meter in the process of composting the decomposed organic fertilizer.

To prevent soil from carrying pathogens, in addition to disinfection by tedding, chemical disinfection should also be performed. The nutrient soil can be evenly sprayed with 400–500 mL of 40% 50-fold-diluted formaldehyde per cubic meter or sprayed with 50% carbendazim wettable powder solution (1,000 kg seedbed soil + (25–30) g carbendazim). Then, fully stirred and covered with plastic film for three to four days to kill the pathogens in the soil that cause blight, seedling blight, and other diseases.

If the seedbed is used for seedling raising, water it thoroughly before sowing. After the water infiltrates into the seedbed, spread a thin layer of sieved fine earth on the border surface and then sow the seeds. The seeds can be mixed with a small amount of fine wet earth before sowing to spread them evenly. After sowing, cover them with 1–2-mm-thick fine earth and add plastic film for warming and moisturizing in the cold season. The ratio of seedbed area to planting area is approximately 1 : 20.

Nursery pots can also be used to raise seedlings. They require fewer seeds for seedling cultivation, and the seedlings grow rapidly without going through a recovery period after being planted in the field, promising a high survival rate and high quality. The nursery pots should not be fully filled but filled to the extent that the nutrient soil is 1–1.5 cm away from the bowl edge to facilitate soil mulching during sowing and after seedling emergence. Watering is required before sowing.

In large-scale production, plug tray substrate is used for seedling raising. For seedlings of different ages, different sizes of plug trays are used: 128-cell plug trays for young seedlings with three to four leaves and 72-cell plug trays for old seedlings with four to five leaves. Lettuce has strict substrate requirements. It prefers moist, loose, well-drained substrate with rich organic matter, with a suitable pH level of 6–6.5. The seedling substrate should be non-toxic and highly porous with stable chemical properties. The substrates commonly used in production are vermiculite, peat, carbonized rice husks, perlite, sand, small gravels, and slag. The substrate is prepared by mixing peat and vermiculite at a ratio of 1 : 1 for seed sowing or at a ratio of 3 : 1 for seedling separation, which is conducive to the transplanting of rooted seedlings and root protection. Generally, two to three seeds are sown per cell, and one strong seedling is retained; 10–15 g of seeds are sown per mu, and more may be needed when sown at high temperatures.

V. Sowing

The sowing time depends on the planting schedule. A proper sowing period is not only related to the local climate and seedling raising conditions but also to the market demand. Generally, the sowing period is 40–50 days (or 60 days earlier for seedling raising in the cold frame) earlier than the planting period. In southern and northern China, which has different climates and uses different cultivation methods, the sowing period is different. The specific sowing date depends on local weather conditions. Sowing should be carried out when it is neither too hot nor too cold, and must not be carried out before the advent of cold air.

The vegetative phase of lettuce is short. Annual production and phased sowing can be realized by cultivation in different protection facilities. A seedling age of 20 days is required if the seeds are sown from April to September, and 30–40 days if sown from October to March. Generally, seedlings should have four to five leaves before planting. Lettuce seeds require light to germinate, so they should not be buried too deep. After being sown, they should be covered with a thin layer of nutrient soil and watered to the extent that they are not exposed.

VI. Seedling Stage Management

After sowing, the seedbed temperature should be maintained at 15–20°C to keep the soil or substrate moist. After 3–4 days, the seedlings will emerge. Then, the temperature should be maintained at 15–18°C during the day and at about 10°C at night, not lower than 5°C at the minimum. The temperature should be strictly controlled (not exceeding 25°C) for sowing from July to August; otherwise, the germination will be slow, and even hot dormancy will occur. When the seedlings grow with two leaves and one cotyledon, the relative water content in the soil should be kept at 75%–80%, and 0.2%–0.3% urea and potassium dihydrogen phosphate solution should be sprayed on the leaf surface while water is sprayed. The seedlings can be separated with a planting spacing of 8 cm × 8 cm when three true leaves emerge and can be planted in the field when four to six leaves emerge.

[Summary]

I. Key Points

Seed germination conditions and sowing and seedling raising techniques.

II. Difficult Points

Determination method of indicators in seed grading standard.

[Skill Training]

Skill Training 3-1: Seedling Raising of Lettuce in Plug Trays

I. Purposes and Requirements

To select the appropriate plug trays for soilless seedling raising according to the lettuce type and seedling age, master the procedures for soilless seedling raising, and accurately analyze and effectively solve the practical problems during soilless seedling raising

II. Planning

1. Preparation of Materials

100 g lettuce seeds; an appropriate amount of vermiculite, perlite, and peat; 3%–5% trisodium phosphate solution; 0.3%–0.5% sodium hypochlorite (calcium hypochlorite) solution.

2. Instruments and Tools

Pregermination chamber, seedling plug trays (72-cell and 128-cell), tweezers, plastic labels, etc.

3. Investigation Plans

(1) Seed selection: Select full, neat, pest-free seeds for future use.

(2) Disinfection of tools and hands: Disinfect with 3%–5% trisodium phosphate solution.

(3) Disinfection of seeds and substrate: Disinfect the selected seeds with 0.1% mercuric chloride solution for about five minutes, and rinse them with clean water three to five times to remove residual disinfectants. Soak the sand-peat substrate with 0.3%–0.5% sodium hypochlorite (calcium hypochlorite) solution for 30 minutes, and then rinse it with clean water several times.

(4) Substrate loading: Mix the perlite and vermiculite in a ratio of 2 : 1 and load them into the trays (1 cm away from the tray edge).

(5) Sowing: Carefully put the seeds into the plug trays with tweezers. Sow one to two seeds per cell, cover with a thin layer of vermiculite substrate after sowing, then scrape the soil flat, gently press it, and water it thoroughly.

(6) Tracking and recording: During seedling raising, monitor and record the growth of seedlings and environmental changes. Timely sort out and archive the seedling records.

III. Implementation

(1) Prepare the plan for raising lettuce seedlings in plug trays.

(2) Conduct seeding experiments and monitor the growth of seedlings.

Skill Training 3-2: Impact of Plug Tray Size on the Quality of Lettuce Seedlings

I. Purposes and Requirements

Small-cell plug trays have a low buffering capacity for water and nutrients, which is likely to restrict the growth of root systems and cause competition for space among overground parts, thus affecting the normal growth of seedlings. In addition, cultivation with different substrates also has different impacts on seedlings. This experiment is designed to allow trainees to learn and master the technique of growing lettuce from plug seedlings and to understand how the plug tray size affects the quality of lettuce seedlings.

II. Planning

(1) Materials: Lettuce seeds, peat, coconut chaff, and perlite.

(2) Instruments and equipment: 50-cell, 72-cell, and 128-cell plug trays, electronic balances, rulers, beakers, gauze, absorbent paper, tweezers, and washbasins.

(3) Facilities: Plastic-board facilities.

III. Implementation

(1) Substrate preparation and loading: Mix the peat and coconut chaff (with a ratio of 1 ∶ 1) well, and load them into different sizes of plug trays. Before loading, measure the cell volume of different sizes of plug trays with a measuring cylinder and make records.

(2) Seed soaking and pre-germination: Soak the lettuce seeds in 55°C warm water. After that, take out the seeds, drain the water, wrap them with gauze, and put them in an incubator for pre-germination.

(3) Sowing: Make a hole in the center of each cell, carefully put the germinated seeds into the holes, and cover them with a layer of substrate with a thickness of 0.5–1.0 cm. Then water them well, and place the plug trays in the plastic-board facility.

(4) Management: After sowing, water the plug tray thoroughly with a sprinkling bucket, observe the plug tray at different times every day, and properly water it according to the moisture in the plug tray and the weather conditions, making sure it is neither too dry nor too wet. Strengthen the management of seedlings after they emerge and their cotyledons spread to cultivate strong seedlings.

(5) Measurement: After seeds grow into seedlings, take 10 seedlings from

three different sizes of plug trays. Measure the plant height, leaf size, and the number of leaves. Then, dry and weigh the seedlings, recording relevant data.

(6) Input the experimental data into Table 3-1 below.

Table 3-1 Records of Growth Status of Lettuce Seedlings in Different Sizes of Plug Trays

	50-cell	72-cell	128-cell
Plant height (cm)			
Leaf size			
Number of leaves (pcs)			
Weight (g)			

Skill Training 3-3: Determination of Lettuce Seed Viability

I. Purposes and Requirements

Seed viability refers to the germination potential of seeds or the vitality of embryos. Seeds without viability are dead seeds that cannot germinate. Master the method for determining the lettuce seed viability, the basic operation skills, and the result calculation method through experiments.

II. Planning

(1) Determination of lettuce seed viability: Triphenyltetrazolium chloride (TTC) method.

(2) Materials: Lettuce seeds.

(3) Instruments and tools: Petri dishes, tweezers, single-sided knives, base plates (for seed cutting), beakers, brown reagent bottles, enameled dishes, and pH test papers.

(4) Preparation of TTC solution: Dissolve 1 g of TTC in 1 L of distilled water or cold boiled water to prepare 0.1% TTC solution. The pH value of the solution should be within 6.5–7.5.

III. Implementation

(1) Soak the lettuce seeds in warm water (30°C) for two to five hours to allow the seeds to fully expand. (Take a portion of fully expanded seeds and boil them in boiling water for three to five minutes to kill them.)

(2) Randomly take 50 seeds, accurately cut them in half along the center of the embryo, and take several dead seeds as control.

(3) Stain the halved seeds in the TTC solution that just immerses them.

(4) Place the TTC-stained seeds in an incubator at a temperature of 30–35°C for heat preservation for 30 minutes.

(5) After that, discard the solution, rinse the seeds with tap water twice or three times, and immediately observe the staining of the embryo to judge whether the seeds are viable. Record the results in Table 3-2.

Table 3-2 Record of Lettuce Seed Viability by Staining

Method	Seed Name	Number of Tested Seeds	Number of Viable Seeds	Number of Inviable Seeds	Percentage of Viable Seeds in Tested Seeds

$$\text{Seed viability (\%)} = \frac{\text{Number of viable seeds}}{\text{Total seeds}} \times 100\%$$

[Extension Tasks]

I. Review Exercises

(1) What are the factors that affect seed germination?

(2) What considerations and specific work need to be done before sowing?

(3) What is the process of lettuce plug seedlings?

II. More to Know

Judgment of Strong Seedlings and Judgment Indicators

Strong seedlings refer to high-quality seedlings with great production potential.

A group of strong seedlings should be pest-and disease-free and grow neatly, healthily, and vigorously. Seedling age is closely associated with seedling quality. Seedling quality depends on seedling age and quantitative traits. Seedling age may refer to a calendar seedling age, a morphological seedling age, or a physiological seedling age. The three are not necessarily the same. The calendar seedling age is the most widely used in production. In general, the growth, development, and harvest of older seedlings are earlier; however, if the seedling age is too old, it is not conducive to the increase of total yield.

The cultivation of strong seedlings is the key to seedling cultivation. At present, the quality of seedlings is generally described by quantitative traits related to seedling growth, such as stem height, stem diameter, number of leaves, leaf area, fresh weight and dry weight of various organs, root volume, flower bud differentiation nodes, and number of flower buds, as well as relative indicators such as stem diameter or stem height, root weight or crown weight, leaf area or root volume, crown fresh weight or crown dry weight, number of leaves, seedling range or seedling height. Stable indicators such as stem diameter, leaf area, seedling dry weight, and crown dry weight, and some relative indicators can reflect the quality of seedlings from different aspects. Under certain conditions, the quality of seedlings can be judged by some indicators alone, such as stem diameter, crown dry weight or number of leaves, seedling range or seedling height. However, incorrect results are possible if these indicators are used to judge the quality of seedlings without restrictions. It is often difficult to comprehensively and accurately judge the quality of seedlings with such indicators. Physiological indicators such as photosynthetic intensity, respiration intensity, root activity, root water absorption, carbon and nitrogen content and their ratio, and the content of sugar, protein, auxin, and nucleic acid reflect the status and quality of seedlings to some extent and can be used for studying the quality of seedlings and even determining strong seedling indicators. However, they should not be directly used as strong seedling indicators because they are not necessarily accurate and well-rounded and there may be some problems with the application. To this end, some equations that can reflect the growth of seedlings are proposed, such as:

$$\text{Strong seedling indicator} = \frac{\text{Stem diameter}}{\text{Plant height}} \times \text{Seedling dry weight (or crown dry weight)}$$

$$\text{Strong seedling indicator} = \frac{\text{Seedling range}}{\text{Seedling height}} \times \text{Number of leaves}$$

These equations not only increase the predictability of early yield but also are very easy to apply.

III. Case Sharing

Outstanding Achievements of Shandong Province in Seedling Production

Shouguang New Century Seedling Co., Ltd., established in 2001, has achieved annual production of more than 30 million seedlings after nearly 10 years of efforts and has taken the lead in the research and development of grafting of tomato, eggplant, and chili (sweet pepper). Relying on the commercial vegetable varieties and the seedling promotion service system, the company has improved market competitiveness and provided farmers with better services.

Due to the complex composition of the seedling substrate and the lack of peat (one of the main components), Jinan Luqing Institute of Horticulture has formulated "standards for fruit and vegetable bases", which has been approved in the expert review organized by Shandong Quality and Technical Supervision Administration and will be issued and implemented as provincial-level local standards. The company has become one of the largest seedling substrate manufacturers in China, and its products have been sold to Shandong, Hebei, Ningxia, Guangxi, and other provinces and autonomous regions.

Significant progress has been made in the research of vegetable seedling substrate, nutrient solution formula, seedling environment conditioning, seedling disease control, and seedling standards in Shandong Province. Through joint efforts of scientific research institutions, colleges and universities, and seedling production companies, the research and development platform has been preliminarily established. The technical procedures for intensive factory-based seedling production of watermelon, cucumber, tomato, eggplant, and chili (sweet pepper) have been formulated. They will be issued and implemented as provincial-level local standards. The scientific and technological progress in seedling production provides important technical support for the development of intensive factory-based seedling production in the province.

Module 4 Storage and Transportation of Lettuce Seedlings

[Learning Objectives]

I. Knowledge

(1) Understand the preparations for the transportation of lettuce and be able to develop a detailed transportation plan;

(2) Understand the technical measures to prevent quality degradation of seedlings during transportation and storage.

II. Skills

(1) Learn to retrieve relevant information;

(2) Be able to develop a detailed transportation plan for lettuce seedlings.

[Preparation]

I. Required Resources

(1) A paper library and a periodicals reading room;

(2) A digital reading room and an electronic resource library;

(3) A multimedia classroom.

II. Background Knowledge

(1) Master the basic methods of literature review;

(2) Understand the relevant knowledge of lettuce botany;

(3) Understand the damage to lettuce seedlings during transportation.

[Learning Tasks]

Task 1 Preparations for Transportation

I. Development of Transportation Plan

Make a plan before the transportation of seedlings, listen to the weather forecast, and transport seedlings in good weather conditions to reduce the loss of seedlings. Buyers should prepare for planting to ensure that seedlings can be planted immediately upon arrival to improve the survival rate.

II. Treatment of Seedlings Before Transportation

Seedlings should be treated with agents for fresh-keeping before transportation to prevent excessive water loss and declined root system activity during long-distance transportation and enhance the recovery of seedlings. For example, the treatment with 300 μg/L ethephon solution can promote the recovery and rooting of seedlings after planting and improve the yield; the treatment with 400–600-fold-diluted 0.5% amino-oligosaccharide aqueous solution can improve the quality of seedlings by 6.1%–17.6% compared with the control after storage and transportation for three to seven days; the compound vegetable seedling preservative with fulvic acid as the main component can be sprayed once one day before storage and transportation, which is easy to use and comes with low costs and good effects.

III. Root Protection of Seedlings

Plug seedlings may be transported in trays, which can protect the root system well, but are only suitable for small freight volumes. If seedlings are not transported in trays, they should be densely arranged to prevent the root system from loosening due to substrate loss.

Task 2 Packaging and Transportation Tools

I. Packaging of Seedlings

After lettuce seedlings are cultivated, they should be packaged and transported promptly. Cartons, wooden boxes and plastic boxes can be used for the packaging of seedlings, and corresponding packaging containers should be selected according to the transportation distance. The containers should be of enough strength to withstand certain pressures and bumps during transportation. In long-distance transportation, each container should not hold too many seedlings. In loading, it is necessary to make full use of the truck space while leaving a certain gap between containers to prevent the damage of heat from the respiration of seedlings. In order to improve the freight volume of seedlings per unit space of the truck, the cultivated seedlings should be removed from the plug trays and placed flat in the containers. In the packing process, care should be taken not to destroy the root systems of seedlings so as not to affect their recovery after planting.

II. Transportation of Seedlings

The main means of transportation in China are trains and trucks. For stability during long-distance transportation, trains are the most appropriate means and should be selected to reduce the losses during transportation. However, train transportation has some disadvantages, such as long-time transportation, repeated loading or unloading operations, and the inability to arrive directly at the planting base. In the actual production, in order to ensure the rapid and safe transportation of seedlings to the planting base, trucks are generally used to reduce the repeated loading/unloading operations. Since there are certain requirements for temperature and ventilation during the transportation of seedlings, trucks equipped with air conditioners are the first choice, but the transportation cost is high. If truck transportation is adopted, heat preservation should be considered in winter to prevent frost damage to seedlings. During the transportation, the plug seedlings should be transported in seedling boxes that are cheap and not easy to crush.

III. Treatment during Transportation

Fresh-keeping techniques for vegetable seedlings during transportation are crucial for the recovery of seedlings after planting. For short-distance or short-time transportation, certain humidity and appropriate temperature are required. For long-distance transportation, appropriate nutrition should be provided for seedlings during transportation. For example, the nutrient solution containing nitrogen, phosphorus, potassium, and other nutrient elements can be sprayed on leaf surfaces, and certain temperature and humidity should be maintained at the same time. Additionally, to avoid the degradation of seedlings resulting from long-time respiration under moist and anoxic conditions, ventilation holes should be made on packaging containers.

Task 3 Techniques to Maintain the Quality of Seedlings during Storage and Transportation

Seedlings are raised annually according to the production needs, and there is almost no choice for weather. In other words, the atmospheric environment for seedling transportation cannot be changed. Therefore, only the specific conditions of transportation and storage and the microenvironment of seedlings can be changed to keep them fresh. Pay attention to the following when taking technical measures to prevent the declined quality of seedlings during transportation and storage.

I. Prevent Seedlings from Frost Damage

In northern China where the temperature is very low in winter, the following preventive measures should be taken to prevent seedlings from freezing or cold damage.

1. Cold Hardening

The seedbed temperature should be gradually reduced three to five days before seedling transportation for cold hardening. Through temperature reduction and water control, seedlings undergo the following changes to enhance their low-temperature resistance: Reduced growth rate, increased accumulation of photosynthates, remarkable rise of fiber and sugar content, a significant increase of leaf tissue, declined water content, thickened epidermis, and increased stomatal resistance. However, excessive cold hardening and water control are not recommended to avoid a decline in the quality of seedlings. In addition, the time of cold hardening should be planned before seedling raising to ensure seedlings are old enough.

2. Spraying Low-temperature Protection Agent

Spraying 1% low-temperature protection solution twice or three times before transportation can enhance the low-temperature resistance of seedlings.

3. Selecting a Better Packaging Method

For the transportation of seedlings in winter, it is advisable to package the seedlings without plug trays to prevent them from frost damage. The seedlings should be packaged with bare roots, i.e. they should be taken out of the plug trays and placed flat in the packing boxes with all sides tightly wrapped with plastic films or other insulation materials to prevent seedlings from cold wind damage.

4. Providing Heat Preservation Covering

After loading into vehicles, the packing boxes should be tightly covered with insulation materials such as quilts on the top and four sides and fixed with ropes to prevent the quilts from being blown away by wind during transportation.

II. Prevent Seedlings from Heat Damage

During hot summers, cooling measures should be taken during transportation to prevent seedlings from heat damage.

1. Avoiding High-temperature Packing

Before storage and transportation, the seedlings are in a vigorous growth stage. Under normal growth conditions, the photosynthetic intensity is high and the respiration is strong. Once they are removed from the normal growth environment (light source and supply of water and nutrients are cut off), the photosynthesis is blocked, but the respiration is still ongoing. Therefore, to reduce the respiration-based consumption of seedlings, it is necessary to appropriately control both the temperature during storage and transportation and the temperature of seedlings during packing. Packing at high temperatures should be avoided where possible so as not to bring "field heat" into the packing box or increase the respiration-based consumption of seedlings, which may result in the decline of seedling quality.

2. Spraying Seedling Preservative

Seedling preservatives can prevent diseases, reduce water evapotranspiration and improve plant immunity. Spraying seedling preservatives at the specified concentration one day before seedling storage and transportation, such as 0.04% folic acid compound preservatives, can prevent diseases, reduce water evapotranspiration and improve immunity to effectively keep the seedlings fresh and significantly enhance their quality.

3. Increasing Humidity in Seedling Packing Box

There is no water supply during the transportation and storage of seedlings. Watering or water spraying is recommended before packing to reduce the temperature, increase the air humidity in the box, and reduce the water evapotranspiration of seedlings, so as to keep them fresh. If the atmospheric temperature is high and cannot be regulated on trucks, measures should be taken to retain water in the root microenvironment (such as wrapping the root system with moisture-resistant materials) to prevent the seedlings from withering.

4. Transporting at Night

As the temperature is relatively low at night, nighttime transportation can avoid the damage to seedlings that would otherwise be caused during daytime transportation. Especially in the hot summer, the diurnal temperature difference may be up to 15–20°C. Therefore, seedlings should be transported at night wherever possible. Additionally, if seedlings are transported at night, they can arrive at the destination the next morning, allowing them to be planted in a timely manner and improving the survival rate.

III. Prevent Seedlings from "Air Drying"

"Air drying" is a major disaster in seedling transportation. During transportation, due to strong winds, the boundary layer of leaves becomes thinner or even disappears, the resistance decreases, and the transpiration accelerates, thus resulting in water loss of seedlings. Therefore, measures must be taken to prevent seedlings from being "air-dried" during transportation.

1. Using a Water-retaining Agent

A water-retaining agent, also known as super water absorbent, moisturizer, or super water-absorbent resin, is an organic polymer. It can absorb hundreds or even thousands of times the weight of water and then slowly release the water to crops, which has a significant impact on the physiological activity of seedlings. The water-retaining agent is effective in improving stomatal status, photosynthetic pigment, root activity, and other physiological indicators.

2. Spraying Plant Growth Regulator

If seedlings are cultivated in hot summer, proper cooling equipment must be provided; otherwise, seedlings are prone to excessive growth. Despite having a high water content, excessively growing seedlings may be withered due to water loss, underdeveloped mechanical tissue, and a protective layer of leaves. According to seedling raising requirements, 2,000–2,500-fold-diluted cycocel or 500–1,000-fold-diluted paclobutrazol should be sprayed during cultivation to keep seedlings strong and reduce their water loss during storage and transportation.

3. Using an Antitranspirant

Antitranspirants (such as fulvic acid) can reduce the stomata of leaves and water transpiration, thus improving the drought resistance of crops. Spraying fulvic acid before storage and transportation can preserve water and resist drought, thus improving the quality of seedlings during storage and transportation.

4. Supplementing Water and Taking Windproof Measures

If seedlings are to be cultivated in summer, they should be well watered before transportation. To prevent the impact of wind and drought, it is necessary to cover the cargo bed to minimize the air flow inside the truck. Ventilation holes should be made on each packing box, and a certain gap between packing boxes should be maintained to prevent the damage of respiration heat to seedlings.

IV. Take Root Protection Measures

Root protection measures should be taken to reduce root system damage during lifting, transportation, and planting of seedlings.

1. Regulating Root Structure of Seedlings

Since the regeneration ability of the root system of seedlings is strong, seedlings can be separated to thin the root system, thus reducing the root system damage during transportation and planting and improving the survival rate.

2. Preventing Root System Damage

Using nursery pots, nutrient soil blocks, and other root protection tools can effectively prevent root system damage and accelerate seedling recovery after planting.

3. Adopting Soilless Seedling Raising Technique

The root systems of seedlings cultivated soilless are developed and can be transported in the trays so as not to damage the root system of seedlings.

4. Controlling Seedling Age

At an older age, seedlings are easier to suffer root system damage when lifted and transplanted, making it difficult for the root systems to recover. In addition, it is less convenient to transport older seedlings. Therefore, the seedling age should be appropriately controlled, and seedlings should be transported at a young age.

[Summary]

I. Key Points

(1) Master the hazards encountered during the lettuce seedling transportation and the effective measures to be taken.

(2) Learn how to prepare for lettuce seedling transportation.

II. Difficult Points

Be able to develop a detailed lettuce seedling transportation plan.

[Skill Training]

Skill Training 4-1: Investigation on Lettuce Seedling Transportation

I. Purposes and Requirements

Investigate the lettuce seedling transportation in a specific region through network inquiry, field investigation, and other methods, and make an objective evaluation and give reasonable suggestions on the lettuce seedling transportation in this region with what you have learned.

II. Planning

1. Field Investigation

Use measuring tools such as tape measures and steel tape measures, as well as recording tools or equipment such as pencils, rulers, notebooks, and tablet computers.

2. Investigation Plans

(1) Investigate the local transportation means and destinations for lettuce seedlings.

(2) Measure and record the specification, performance, and characteristics of cartons and other containers used for seedling transportation.

(3) Record the environmental control parameters during transportation and the application of antitranspirants or other measures.

III. Implementation

(1) Work in teams. Each team develops an investigation plan for a region.

(2) Design investigation forms, conduct field investigations, and keep records.

(3) Develop the investigation report and analyze the current transportation situation of lettuce seedlings in each region.

(4) Prepare PPT slides based on the investigation report and make a presentation in class.

[Extension Tasks]

I. Review Exercises

(1) What are the common seedling preservatives?

(2) What are the hazards during seedling transportation?

(3) Which is the main transportation mode for lettuce seedlings?

II. Case Sharing

Current Situation of Long–distance Seedling Transportation

There are many modern seedling companies in the United States, the Netherlands, France, Italy, and other countries in the world. These companies have corresponding seedling factories specializing in the sale (including transportation) of vegetable seedlings in their own countries and abroad. For example, the Von Den Beckenlom Seedling Breeding Company in France has a glass greenhouse of 12 hm^2 and produces 250 million seedlings per year, of which 40% are exported, with an annual average output of 2.5 million seedlings per worker. In the Netherlands, almost 99% of greenhouse vegetable growers buy seedlings. In France and Italy, where seedling raising started late, 40%–50% of seedlings are purchased. A large number of vegetable seedlings grown in the Southern United States are transported to the north every year. For example, hundreds of millions of tomato seedlings are transported from Georgia to the middle or north of the United States for processing and cultivation. The development of modern industry and the modern transportation sector has made the long-distance transportation of a large number of vegetable seedlings possible. It takes only a few days to reach places that are thousands of kilometers away.

China began raising and long-distance transportation of seedlings for field planting very early, and it differs from foreign factories where they have been raising and transporting seedlings on a large scale. For example, most onion seedlings planted in Shenyang and Fushun are transported from the Hebei Province or Haicheng and Anshan. Transfers between neighboring cities, especially between townships within a city, have occurred over the years, but most are not preplanned. On the night of May 5, 1985, the seedlings of cucumber, tomato, pepper, and eggplant cultivated with the nutrient solution were transported from Ulanhot, Inner Mongolia, to Aershan, Greater Khingan Range for planting in protected fields at very low geothermal temperature. The survival rate was 92% for various seedlings, including cucumber, tomato, pepper, and eggplant. In 1978, the seedlings of cucumber, pepper, and eggplant were transported by truck from western Fushun to Xinbin, 160 km away for planting, and the survival rate was over 95%. It is very common to raise seedlings in different townships within a county.

In recent years, vegetable seedling raising centers have emerged all over China. Many vegetable seedlings in Shanghai, Jiangsu, etc. are cultivated by seedling raising centers. There are several large-scale modern vegetable seedling raising facilities in Beijing, where seedlings sold as commodities and transported to different places. As the seedlings are cultivated in seedling raising centers in a centralized manner, modern new seedling raising techniques, advanced equipment, and excellent seeds can be used to facilitate standardization. Therefore, the seedlings have strong vitality, making it possible for further cultivation in different places. Temperature varies greatly between north China and south China in spring. While seedlings can be cultivated in an open field or a simply protected field in region A, they can only be raised in a warming greenhouse in region B. It is feasible to utilize this difference to transport the seedlings raised in region A to region B. Especially in two regions close to each other with large temperature differences, the economic benefits can be higher. As mentioned earlier, Ulanhot and Aershan are less than 300 km away, but the climate in the two cities is quite different (the latter is one more month colder than the former). The high cost of raising seedlings in region B, especially in protected fields, makes it easier to promote cross-region seedling cultivation, which costs less and is convenient. The cost of tomato seedlings in the greenhouse in Fushun city is about CNY 600 per mu. These seedlings must be cultivated in a warming greenhouse during the sprout and seedling phases and in a solar greenhouse with better insulation performance in the vegetative phase. Cross-region seedling raising may be considered when the total cost of seedling raising and transportation is less than CNY 600. Definitely, it is also necessary to conduct experimental research and address a series of problems in the cross-region transportation of seedlings, such as problems related to planning and management, transportation means, and packaging containers.

Module 5 Lettuce Cultivation Mode and Technology

[Learning Objectives]

I. Knowledge

(1) Understand the cropping pattern of open cultivation;

(2) Master the cropping pattern for lettuce cultivation in facilities;

(3) Understand the types of lettuce soilless cultivation;

(4) Understand the characteristics of Nutrient Film Technique (NFT) and be able to carry out production with NFT;

(5) Understand the characteristics of Deep Flow Technique (DFT) and be able to carry out production with DFT;

(6) Understand the characteristics of aeroponics and be able to carry out production with aeroponics.

II. Skills

(1) Learn to retrieve relevant information;

(2) Be familiar with the concept, characteristics, and application of DFT, NFT, and aeroponics.

[Preparation]

I. Required Resources

(1) Understand the characteristics of lettuce cultivation through the Internet;

(2) Observe cultivation with DFT and NFT in the Agricultural Expo Garden;

(3) Observe vegetable cultivation in Shouguang, Shandong Province, if possible;

(4) Standardize operation at the lettuce production base.

II. Background Knowledge

(1) Master the basic knowledge of DFT, NFT, and aeroponic facilities;

(2) Master the basic knowledge of literature review;

(3) Understand the basic principles of some DFT, NFT, and aeroponic facilities;

(4) Understand agricultural meteorology and agricultural regulations.

[Learning Tasks]

Task 1 Open Cultivation of Lettuce

Lettuce is a cold-hardy vegetable, and its open cultivation is mainly carried out in spring and autumn.

I. Cultivation Period

For spring cultivation in North China, lettuce is generally sown from February to March and harvested from May to June. For autumn cultivation, it is sown from late July to late August and harvested from October to November. For spring cultivation in the Yangtze River basin, the lettuce is sown in the plastic greenhouse from the end of January to early February and transplanted in early March; while for open cultivation, it is sown in early March and transplanted in early April. In autumn, the lettuce is sown in August and harvested from September to October. In South China, the lettuce can be sown from October to following February, and harvested and sold from September to the following April.

II. Variety Selection

Early-maturing varieties should be selected for open cultivation in spring. Heat-resistant or looseleaf varieties should be selected for cultivation in summer, while low-temperature varieties with strict temperature requirements (cold climate during heading stage) should be selected for cultivation in autumn.

III. Sowing and Seedling Raising

When the average monthly temperature is higher than 10°C, seedlings can be raised in the open field; when it is lower than 10°C, facilities are needed to raise them. For sowing, seeds newly harvested should be selected. Considering the difficulty in seed germination under high temperatures, the seeds should be soaked in warm water at 15–20°C, scrubbed, and taken out to store in a refrigerator at 5°C for 24 hours before sowing. It is also feasible to soak the seeds in 5 μL/L gibberellin solution for 6–7 hours, so as to increase the germination ratio. Broadcast sowing or sowing in plug trays are both acceptable. To promote seedling emergence, seeds can be covered by mulch film for warming and moisturizing if sown in winter and spring, or covered with shade nets or straw for moisturizing and cooling if sown in summer and autumn.

After emergence, the temperature should be controlled within 16–20°C during the day and around 10°C at night. When seedlings grow to have 2–3 true leaves, they may be separated and planted in nursery pots or seedbeds. After that, they should be watered and covered immediately for heat preservation and moisturization. After recovery, watering should be properly controlled to facilitate rooting and strong growth.

IV. Planting

During deep plowing, 4,000–5,000 kg of fully decomposed organic fertilizer, 20 kg of calcium superphosphate, or 50 kg of ternary compound fertilizer can be applied per mu for cultivation on low or high borders. Seedlings at the age of 30–40 days with 4–5 true leaves can be planted. The root system should be protected where possible during planting, which can be carried out on cloudy days or in the evening. The plants should be graded to achieve growth uniformity on the same border. The seedlings should not be planted too deep but should be planted stably. The method of planting seedlings before watering should be adopted. Lettuce has a long growth period in spring, with many outer leaves, a large lateral spread, late heading, and a high yield. Therefore, a proper planting density is required with a row spacing of 40–45 cm and a plant spacing of 25–35 cm. Autumn lettuce or looseleaf lettuce should be relatively densely planted, with both the row spacing and plant spacing being 25–30 cm. For the lettuce plants that are harvested young for food, the row spacing and plant spacing should be 10 cm×10 cm. The seedlings should be supplied with sufficient water after being planted in the field and should be watered in a timely manner to boost recovery according to the weather conditions.

V. Management After Planting

After recovery, seedlings should be watered and fertilized at the same time, and seedling hardening is not required. Watering should be performed based on soil moisture and growth conditions, generally once every 5–7 days. Due to the low temperature in spring, low-flow watering should be done at a long interval; during the peak growth period, sufficient water should be sprayed to retain soil moisture. After leaf heads are formed, watering should be controlled to prevent head split or rot caused by uneven watering. Watering should be stopped before harvesting to benefit the storage and transportation of harvested leafy heads. Supplemental fertilizer should be applied once during watering 15–20 days after planting: 15–20 kg of ternary compound fertilizer can be applied per mu. Then, the land should be intertilled and loosened to improve soil permeability, boosting the extension of the root system and the growth of outer leaves. To prevent premature senescence of outer leaves and promote the enrichment and plumpness of leaf heads, supplemental fertilizer may be applied again during watering 10–15 days after the first time of supplementing: 10–15 kg of urea can be applied per mu.

VI. Harvest

The harvest period of looseleaf lettuce is not strictly specified and can be determined according to the market and growth conditions. For head lettuce, it is required that the leaves should be plump and bend inward to form a compact leafy head. Harvesting too early may affect yield while harvesting too late may lead to elongated center stalk or even bolting, resulting in a decline in quality. When harvesting, gently press the top of the leafy head by hand to judge its compactness. A proper leafy head should be moderately compact as being too compact may result in cracking of the leaves. When harvesting, the leafy head should be gently cut with a knife rather than being twisted off by hand. Three to four outer leaves should be retained to protect the leafy head.

Task 2 Lettuce Cultivation with Facilities

I. Greenhouse Cultivation of Early Spring Lettuce

1. Variety Selection

Head lettuce should be large-head early-maturing varieties with strong adaptability and resistance to cold, diseases, and bolting to ensure early marketing, such as Imperial, Wins, and Great Lake 659 lettuce. Looseleaf lettuce can choose varieties such as Guangdong Iceberg lettuce, Grand Rapids lettuce, Italian lettuce, soft-tailed lettuce, etc.

2. Land Preparation for Planting

For lettuce cultivation, loose soil with rich organic matter and water and fertilizer retention capacity should be selected. Before planting, 5,000–6,000 kg of decomposed farmyard manure, or 1,500 kg of decomposed organic fertilizer, 40–50 kg of calcium superphosphate, 8–10 kg of potassium chloride, and 20–25 kg of ammonium sulfate should be applied per mu during land preparation. The land should be deeply plowed for 25 cm and leveled to make borders. Generally, lettuce is cultivated on low borders with a width of 1–1.2 m, or on high borders with a width of 0.8–1 m. When the temperature of 10-cm-deep soil layers is stable above 5°C, the seedlings can be transplanted, with a row spacing of 40–45 cm and a plant spacing of 25–35 cm. 5,000–6,000 head lettuce seedlings or 6,000–8,000 looseleaf lettuce seedlings can be planted per mu. For looseleaf lettuce, the plant spacing should be 15 cm × 15 cm, and the planting density should be 25,000–30,000 seedlings per mu. The root system and leaves should not be damaged during the lifting of seedlings. The seedlings should be transplanted with soil, and not planted too deep. Otherwise, it will take a longer time for them to recover. After being planted in the field, the seedlings should be watered in time.

3. Management After Planting

(1) Temperature management: Due to the low temperature in the early stage, multi-layer mulching may be adopted for thermal insulation after transplanting. For head lettuce, from recovery to the period heading, the temperature should be maintained at 15–20°C during the day and not lower than 10°C at night; from the formation to the growth of the leafy head, the temperature should be maintained at 18–20°C during the day and 12–15°C at night. To prolong the supply period, temperature should be maintained at 10–15°C during the day and not lower than 5°C at night.

(2) Fertilizer and water management: If sufficient water is supplied during planting, recovery can be realized in 3–4 days afterwards. Sufficient water should be supplied again 7 days after recovery. Then the land should be intertilled in time to keep the soil moist and loose during the development period. Head lettuce requires a high moisture content. Generally, the seedlings should be watered immediately when the surface soil becomes dry in the seedling stage. The moisture should be appropriately controlled in the tillering stage. The water supply should be increased in the early heading stage. However, the moisture content should not be too high in the late heading stage to avoid head-splitting. Watering should be stopped before harvest to avoid the rotting of the lettuce heads during storage and transportation. As the head lettuce has a long vegetative phase, a large quantity of fertilizer is required; in addition to sufficient base fertilizer, supplemental fertilizer should also be applied during the growth period. In general, supplemental fertilizer is applied for the first time after the seedlings are watered for recovery based on the criterion of 15–20 kg of ternary compound fertilizer or 5–10 kg of ammonium nitrate per mu. During heading, supplemental fertilizer is applied for the second time based on the criterion of 10–15 kg of ternary compound fertilizer or 15–20 kg of ammonium

nitrate per mu. For looseleaf lettuce with a short vegetative phase, foliar fertilizer such as 0.2% potassium dihydrogen phosphate solution is generally applied twice to three times during watering. Moreover, watering should be both timely and sufficient, and waterlogging should be avoided.

(3) Disease and pest control: Lettuce has a special aroma and is thus resistant to diseases and pests. *Sclerotinia* disease and tip burn are the two main diseases to be controlled in production.

4. Harvest

Head lettuce varies in maturity period and may be harvested while matured according to the characteristics of lettuce varieties. Generally, it is harvested 50–60 days after being planted in the field. The harvesting criterion is when the leafy head becomes moderately compact. Early harvesting affects yield, while harvesting too late will lead to an elongated center stalk and loose and easy-to-rot leafy head, resulting in a decline in quality. For looseleaf lettuce, there is no strict harvest criterion and the harvesting time can be flexibly adjusted according to market demand and lettuce growth status. Generally, they can be harvested 40–50 days after being planted in the field.

II. Delayed Cultivation of Greenhouse Lettuce in Autumn

1. Variety Selection

The delayed cultivation of greenhouse lettuce in autumn is conducted from mid-August to mid-September when mid-maturing varieties with heat and disease resistance are generally selected. After mid-September, large-head varieties with cold resistance may be selected.

2. Land Preparation for Planting

The land should be cleaned by removing broken branches and weeds from the greenhouse. Then, fertilizer should be evenly spread based on a criterion of 6,000–7,000 kg of decomposed organic fertilizer or 450–500 kg of organic compound fertilizer per mu. After that, the land should be deeply plowed and harrowed to loosen the soil and mix it well with the fertilizer. On this basis, small, high borders with flat surfaces are required. Head lettuce may be planted on borders with a width of 0.9–1.2

m, in three rows per border with a plant spacing of 30–35 cm. About 3,300 plants can be cultivated per mu. Looseleaf lettuce may be planted more densely.

3. Management after Planting

(1) Temperature management: After planting, from recovery to the period before heading, the temperature should be maintained at 15–20°C during daytime and not lower than 10°C at night. From the start of head formation to the growth of the leafy head, the temperature should be maintained at about 20°C during daytime and 15–18°C at night. To prolong the supply period, the temperature should be appropriately lowered to a range of 10–15°C during daytime and not lower than 5°C at night. From the coldest December to next January, thermal insulation measures should be taken by covering the head lettuce with multi-layer plastic films or non-woven fabrics.

(2) Fertilizer and water management: The seedlings should be sufficiently watered before planting, immediately watered after planting, and watered again seven days after planting for recovery. The soil should be kept moist from recovery until the lettuce leaves seal the ridges. After watering, the land should be intertilled once or twice. To prevent the occurrence and spread of diseases, borders must not be directly watered after the lettuce leaves seal the ridges and irrigation of the ridges between them is allowed. Watering should be stopped two weeks before harvest. To guarantee the nutrient demand of head lettuce, supplemental fertilizer should be applied once or twice according to soil texture, application of base fertilizer, and growth of plants. Two weeks after planting, i.e. when the plants have five to six leaves, supplemental fertilizer may be applied for the first time based on 10 kg of ternary compound fertilizer per mu; supplemental fertilizer may be applied for the second time at the beginning of heading based on 10 kg of ternary compound fertilizer per mu. Foliar fertilizers such as 0.2% potassium dihydrogen phosphate solution can be sprayed on the leaves of looseleaf lettuce twice to three times.

(3) Disease and pest control: In this stage, prevailing diseases and pests are downy mildew, gray mold, *Sclerotinia* disease, virus disease, aphids, beet armyworms, cotton leafworms, and oriental tobacco budworms that should be prevented and controlled.

4. Harvest

The lettuce plants to be harvested should have compact heads and be free from diseases or pests with typical characteristics of the variety. To avoid the impact of low temperatures, the plants satisfying the harvest criteria should be harvested in a timely manner.

III. Cultivation of Lettuce in Solar Greenhouse

With good thermal insulation, the solar greenhouse can meet the temperature conditions required for lettuce growth. Lettuces can be cultivated in a solar greenhouse in all four seasons, hence a variety of lettuces, namely autumn-to-winter crops, overwintering crops, and winter-to-spring crops.

1. Variety Selection

Head lettuces with heat and cold resistance, high yield, and high quality can be selected, such as the Imperial and Kaiser lettuce. Looseleaf lettuces with cold resistance, strong disease resistance, high yield, good quality, and adaptability in greenhouse cultivation can be selected, such as Great Lake 659, glasshouse lettuce, and butterhead lettuce.

2. Land Preparation for Planting

Base fertilizers should be applied sufficiently in combination with land preparation, based on 5,000 kg of decomposed farmyard manure and 15 kg of ternary compound fertilizer per mu. The land should be deeply plowed and well-harrowed to make borders with a width of 1–1.2 m. Head lettuces should be cultivated at a plant spacing of 25 cm × 35 cm, while looseleaf lettuces can be planted more densely at a plant spacing of 20 cm × 30 cm. During planting, seedling balls should be leveled with the border surface and be watered sufficiently.

3. Management after Planting

(1) Temperature management: When the external temperature drops to 2°C, the seedlings should be mulched and air may be let in through bottom openings to allow them to adapt to the environment. In case of frost, the seedlings must not be ventilated at night but should be ventilated during the day. The temperature should be maintained at 18–22°C during daytime and 12–14°C at night. Lighting

is increased by using shed plastic film with high light transmittance, adding or removing the straw mat and cleaning the dust on the shed plastic film in a timely manner. The ventilation of the greenhouse should be strengthened. Even when it is cloudy for days, short-time ventilation is required to lower the humidity in the greenhouse and prevent diseases and pests caused by excessive humidity. When the external temperature is lower than 0°C, the straw mat or thermal insulation quilt should be used to strengthen thermal insulation.

(2) Fertilizer and water management: The seedlings should be watered for the second time 7–10 days after recovery. During this process, supplemental fertilizer should be applied based on 20 kg of ternary compound fertilizer or 5–7 kg of urea per mu. The land should be intertilled timely, but not too deep since the lettuce has a shallow root system. The hardening of seedlings should be performed during intertillage but should end when the inner leaves begin to turn green. At this time, the growth of seedlings should be “promoted” rather than “controlled”. Generally, the seedlings are watered once every 5–7 days to keep the relative water content of the border soil at 60%–70%. During watering, supplemental fertilizer may be applied based on about 10 kg of ternary compound fertilizer per mu. The plants should be evenly watered to avoid head split or rot. Watering amount should be controlled as temperature decreases, and watering and fertilization should be stopped 10 days before harvesting.

(3) Disease and pest control: The main diseases are downy mildew and soft rot, and the pests are mainly aphids and cabbage moths, which should be prevented and controlled.

4. Harvest

About 50 days after planting, lettuces can be harvested successively according to market demand, but the harvest period should not be too long to avoid head split and decreased marketability. The looseleaf lettuces that can be harvested must be insect-free, disease-free, large in size, and bright in color.

IV. Cultivation of Oversummering Lettuce by Shading

Lettuce prefers cool days in nature and is mainly cultivated with shade nets in summer when the temperature is high and the light is strong.

1. Variety Selection

Varieties with strong heat resistance, disease resistance, and early maturity, such as Olympia, Kaiser, and Imperial, should be selected.

2. Cultivating Strong Seedlings

(1) Sowing: Summer lettuce is generally sown from the end of May to the beginning of June and harvested and sold from August to September, an off-season for vegetables. Since the external temperature is so high during sowing and exceeds the optimal temperature for germination, seed treatment is required to increase the germination rate.

(2) Seedling stage management: Cover the borders with silver-gray aluminum foil shade nets with a shading rate of about 50% for moisturizing and cooling after sowing. Before the emergence of seedling, if the seedbed is dry, water can be sprayed with a sprinkler to increase humidity. If the seeds emerge with their coats on, a layer of fine soil can be spread on the border surface in the evening. When most of the seedlings emerge, a small arched shed can be built with a shade net on it for cooling, preventing rain, and reducing water loss, thus promoting the robust growth of seedlings. When the seedlings grow with 2–3 true leaves, the spacing between them can be maintained within 6–8 cm. The seedlings aged 20–25 days with 4–5 true leaves can be planted.

3. Land Preparation for Planting

One week before planting, the shed should be covered with a shade net. During land preparation, 3,000–4,000 kg of decomposed organic fertilizer and 20–25 kg of ternary compound fertilizer can be applied per mu. The land should be deeply plowed and well-harrowed to make high borders with a width of 0.9–1.2 m or low borders with a width of 1.5 m. Before planting, the seedbed should be watered thoroughly to make it easier to lift seedlings with soil. The seedlings should be transplanted on a sunny evening or a cloudy day, with a plant spacing of (20–25) cm × 30 cm. Such a planting depth can allow the soil balls rather than inner leaves to be buried. The seedlings should be watered immediately after being planted in the field.

4. Management after Planting

(1) Temperature management: When the temperature in the shed rises to 25°C,

the shed should be covered with a shade net in time to prevent high-temperature hazards; when the temperature drops to 20°C in the afternoon, the shade net can be removed to increase internal light.

(2) Fertilizer and water management: Lettuce cultivation requires a lot of water, and the field should be kept moist. After the lettuce grows to have 5–8 new leaves, 10–15 kg of urea should be applied per mu during watering. At the beginning of the heading, the watering amount should be increased, and second-time supplemental fertilization is required during watering based on 15 kg of ternary compound fertilizer per mu. Watering should be done evenly in the evening and stopped 3–4 days before harvesting to facilitate storage and transportation.

(3) Disease and pest control: Common diseases and pests, including virus diseases, downy mildew, soft rot, aphids, and cutworms, should be prevented and controlled.

5. Harvest

The lettuces should be harvested in a timely manner according to the characteristics of the variety. The leafy heads are ready for sale once they are cut off and the external old leaves are removed. Looseleaf lettuces are sold after the root systems together with external old and yellow leaves are removed. In the case of long-distance transportation, the lettuces should be precooled in a refrigerator at 2–8°C for 24 hours and packaged with preservative film before transportation.

Task 3 Lettuce Soilless Culture

I. Concept of Soilless Culture

Soilless culture is a method of growing plants in any nutrient solution or substrate other than natural soil. It provides growth conditions such as moisture and nutrients for crops to grow normally and complete their entire life cycle. In short, soilless culture is the way to grow plants without natural soil. It used to be called solution culture or hydroponics in the early stage as nutrient solutions were used early and long for culture.

As defined by the International Society of Soilless Culture (ISOSC), soilless culture refers to all methods that use balanced nutrient solutions (with or without substrates) rather than natural soil to supply nutrients for plant growth and development, enabling the plants to complete the entire life cycle.

II. Types of Soilless Culture

Soilless culture has a history of more than 140 years since the beginning of early laboratory studies. In the process of moving from laboratory to large-scale commercial production and application, soilless culture has evolved from the original basic technique developed by German scientists Sachs and Knop in the mid-19th century to a wide range of techniques at present. It has been classified by different researchers from different perspectives, some according to the morphology of the substrate, some based on the type of the substrate, and some according to the shape of the facilities used, leading to different results. It is difficult to classify soilless culture in a scientific manner. Currently, it is commonly classified into two types: non-solid substrate culture and solid substrate culture, based on whether there is a solid substrate in the root system growing environment. These two types are further divided according to the types of materials used to fix the root system of plants and the cultivation techniques.

1. Non-solid Substrate Culture

Non-solid substrate culture is also known as solution culture. It refers to the method in which the root system is not fixed in solid substrates but grows in nutrient solution or humid air containing nutrients, and solid substrates are generally not used in the rhizospheric environment except in seedling cultivation. The method is called hydroponics if the root system grows in a nutrient solution and aeroponics if the root system grows in humid air containing nutrients. Hydroponics is further classified according to the depth of the nutrient solution.

(1) Nutrient Film Technique (NFT): NFT (Fig. 5-1) is a technique where the nutrient solution flows at a depth of 1–2 cm. NFT facilities include growing channels, reservoir tanks, nutrient solution circulation systems, and some auxiliary facilities.

Fig. 5-1 Nutrient Film Technique

Advantages:

① Low investment in facilities, easy and convenient construction: The NFT channel is made of lightweight plastic film or spliced with corrugated tiles. Light and simple in structure, it is easy to assemble and disassemble and requires low investment.

② Shallow flowing stream of solution: The root systems of the crop are partially immersed in the shallow stream of nutrient solution and partially exposed to the moisture in the channel. With the circulation of the nutrient solution in the channel, the oxygen demand of the root systems can be better met.

③ Easy for automatic management of production process.

Disadvantages:

① Although the NFT facilities require less investment and are easy to construct, subsequent investment and frequent maintenance are needed owing to their poor durability.

② The shallow stream of solution and intermittent supply in NFT facilities can better satisfy the oxygen demand of the root system. However, inadequate stability of the rhizospheric environment requires highly qualified management persons and higher-performance equipment.

③ To streamline management, automatic control devices are essential, resulting in increased equipment and investment, which restrains promotion.

④ Since the NFT system is a closed circulatory system, a disease, once it breaks out in the root system, is more likely to spread throughout the whole system. Therefore, strict requirements must be followed before use in the cleaning and disinfection of facilities.

(2) Deep Flow Technique (DFT): DFT (Fig. 5-2) is a culture technique that requires the depth of nutrient solution to be above 5–6 cm. Given the stable liquid temperature and the fact it is not affected by power outages and water cutoffs, this technique may be promoted in subtropical and tropical areas.

Fig. 5-2 Deep Flow Technique

Cultural characteristics of DFT have three aspects. ①Deep growing channels and nutrient solution: With an adequate amount of nutrient solution in the channel, the plants will witness little change in the composition, concentration, pH level, moisture, and temperature of the nutrient solution during growth, which is good for maintaining a stable environment for the growth of root systems. ②Circulating nutrient solution: The circulation of the nutrient solution can increase the concentration of dissolved oxygen in the nutrient solution, eliminate the locally deposited harmful metabolites and supplement for nutrient deficiency in the root zone, and promote the re-dissolution of nutrients that have become ineffective due to precipitation.③Partially hanging of plants in the nutrient solution: The root neck is hung above the liquid level, which can avoid decay or even death of the plants caused by hypoxia during growth.

Advantages:

① Culture by hanging: The plants are fixed on a plate, with part of the root system inserted into the nutrient solution and part of the root system aerated in the gap between the solution surface and the plate, making them semi-hydroponic and semi-aerial. This makes it easier to balance the water and air supply of the root system.

② Deep stream of solution: The root system stretches into the deep flow, ensuring an adequate supply of solution for a single plant. The large volume and deep stream of solution contribute to little sudden change in the concentration of the nutrient solution (including total salt and nutrients), dissolved oxygen, pH level, temperature, and moisture, providing a stable environment for the growth of the root system. This is the outstanding advantage of deep-flow hydroponics.

③ Circulating nutrient solution: The circulation of the nutrient solution can increase the content of dissolved oxygen in the nutrient solution, eliminate locally deposited harmful metabolites on the root surface (the most obvious impact is the pH level), and liminate the difference in the concentration of solutions at the root and in other places to allow nutrients to be timely sent to the root surface to fully meet the growing needs of the plant and promote the re-dissolution of nutrients that have become ineffective due to precipitation to prevent the occurrence of nutrient deficiency. Therefore, even if it is to cultivate marsh plants or plants that can form oxygen-conducting tissue, it is necessary to circulate the nutrient solution.

④ There are many types of crops suitable for cultivation with this technique. Except for root and tuber crops, almost all fruit and leafy vegetables can be cultivated with it.

Disadvantages:

The initial investment is ralatively large.

(3) Floating Capillary Hydroponics (FCH): FCH (Fig. 5-3) is a hydroponic technique in which plants grow on moist non-woven fabric placed on a piece of foam plastic floating in a nutrient solution that is about 5–6 cm deep. This technique is widely used for the cultivation of vegetables such as tomato, cucumber, cantaloupe, and head lettuce. The FCH facility consists of a growing channel, an underground reservoir tank, circulation pipes, and a control system. Except for the growing channel, the other three parts are basically the same as those of the NFT facility.

Fig. 5-3 Floating Capillary Hydroponics

Advantages:

① This technique works with a floating board and capillary mat in the cultivation bed to create an oxygen-rich environment for cultivating roots in moisture, balancing the water and air supply.

② A large amount of nutrient solution is stored in a horizontal cultivation bed to ensure an adequate and stable supply of fertilizer and water during power outages. A heating coil is used to warm the cultivation bed in winter, and deep-well water is used for cooling in summer to ensure stable rhizospheric temperature and humidity.

③ This technique features low investment in equipment, low power consumption, and convenient installation, operation, and maintenance. The facility is composed of a cultivation bed, a reservoir tank, a circulatory system, and a control system. The nutrient solution is transferred into the cultivation bed via a pipe (with an air mixer) by a timer-controlled water pump and then returns to the reservoir tank. The closed-loop circulation of nutrient solutions is rarely affected by external environmental conditions, ensuring little change in the rhizospheric temperature and humidity, which is suitable for the growth of various plants.

(4) Aeroponics

Aeroponics is a soilless cultivation method in which the plant root system is suspended in a container, and the nutrient solution is transferred by a water pump and sprayed through the inner nozzle of the container to the root system surface at intervals so as to meet the needs of the plant growth. Aeroponics is further classified into spray culture and semi-spray culture. The former is aeroponics, and the latter is a method in which part of the root system is in the nutrient solution, while the other part grows in the fine mist of the solution.

Advantages:

① This technique can well solve the problem of oxygen supply to the root system, and almost no poor growth will occur due to hypoxia.

② With a high utilization rate of nutrients and water, a rapid and effective supply of nutrients, and full use of the space in the greenhouse, the planting quantity and yield per unit area can be improved. The utilization rate of greenhouse space is twice to three times higher than that of traditional plane culture.

③ This technique makes it easier to realize the automation of culture management.

Disadvantages:

① Large investment in production equipment and high reliability of equipment is required. Otherwise, it is easy to cause such problems as nozzle blockage, uneven spray, and excessively large droplets.

② There are high requirements on the management technology as the concentration and composition of the nutrient solution are prone to large variations.

③ The spray device will fail in case of a short-time power outage, which can easily cause damage to the plant.

④ As a closed system, diseases in the root system can spread easily if not controlled properly.

2. Solid Substrate Culture (SSC)

SSC refers to the method of growing plants in an environment with a wide variety of natural or synthetic materials as substrates to fix the root system of such plants and maintain their supply of nutrients and oxygen. In this way, the plants can grow in a rhizospheric environment with a stable and coordinated supply of water, air, and fertilizer to normally complete their life cycle. Substrate culture can well balance the supply of water and air in the rhizospheric environment and requires less investment, which is convenient for production with local resources. Substrate culture is further classified into two types: organic and inorganic substrate culture.

(1) Organic substrate culture mainly refers to peat culture, sawdust culture, straw substrate culture, culture with rice husk ash, and culture with coconut coir.

(2) Inorganic substrate culture mainly refers to sand culture, perlite culture, gravel culture, rockwool culture, vermiculite culture, culture with plastic foam, and ceramsite culture. In general, the substrates are put into plastic bags or growing channels for cultivating crops, which have a certain buffer function and are safe for application.

III. Advantages of Soilless Culture

1. High Yield, High Quality, and High Commodity Rate

As in soilless culture, the growth needs of crops can be adjusted artificially where possible, and the yield per unit is higher than that through soil cultivation. In addition, soilless culture can be applied for production all year round, with a high annual yield. Vegetables cultivated in this way are large in size and of excellent quality. Soilless culture can reportedly increase the vitamin C content in tomatoes by 30%.

2. Improved Land and Space Utilization

Through soilless culture, the land that is not suitable for cultivating crops, such as saline-alkali land, barren mountainous land, wasteland, and island, can be made full use of. Particularly, it can help solve the problem of increasing diseases and pests in the greenhouse due to continuous cropping for years, as well as the aggravated problem of secondary salinization. Moreover, the space of the greenhouse can be utilized to raise the yield per unit and increase the income of farmers.

3. Time- and Labor-saving and High Utilization Rate of Resources

Soilless culture technology, once put into use, can save the tedious labor such as intertillage, fertilization, and weeding, bringing high yield, high output, and high productivity.

IV. Precautions in Soilless Culture

1. Source, Disposal, and Disinfection of Substrates

The substrates used currently come in a wide variety of forms, from different sources to different physical and chemical properties. It is required to mix them at a ratio and disinfect them during use, and their disposal and disinfection after use are also very troublesome. All these restrict the application of substrate culture to some extent.

2. Prevention and Control of Diseases

Hydroponics features a rapid spread of pathogens due to the circulation of the nutrient solution. Once infected with pathogens, all plants may be at risk of being infected. Therefore, more research should be conducted on the nutrient solution disinfection equipment and effective prevention and control agents.

3. Regulation of Greenhouse Environment

At present, common plastic greenhouses and solar greenhouses are mainly used in China for the soilless culture of vegetables, without corresponding regulating equipment, resulting in a low level of greenhouse environment regulation. The cost of introducing equipment from advanced countries is also high. Therefore, the current top priority is to conduct research and develop a proper soilless culture system for cultivating vegetables in China as soon as possible through introduction, digestion, and absorption.

4. Breeding of Special Varieties

Currently, there is almost no vegetable variety specifically suitable for soilless culture. Due to the particularity of soilless culture, there is an urgent need for special high-quality and high-yield varieties that are resistant to low temperatures and diseases in the root system and are adaptable to weak light.

5. High Requirements for Growers in Soilless Culture

Growers should master agricultural production technologies, physiological and biochemical knowledge about vegetables, and mechano-electronic technologies. At present, few technicians have master these techniques and are engaged in the soilless culture in China. Therefore, agricultural colleges and universities and some agricultural research institutions should strengthen the training of professional technicians and promote relevant technologies to improve production efficiency.

V. Essentials for Soilless Culture of Lettuce

The substrate culture of lettuce is relatively simple. In the following text, hydroponics is taken as an example to introduce it.

1. Variety Selection and Cropping Pattern

Early-maturing lettuce varieties with heat and bolting resistance are suitable for hydroponics, such as Butterhead lettuce No.1, Imperial lettuce, and Italian lettuce with bolting resistance. Due to the short growth cycle of looseleaf lettuce, there is no strict criterion for harvest. Therefore, in facilities with good environmental conditions, sowing can be arranged all year round (10–20 times per year). The sowing times of the head lettuce can be properly reduced since it has a longer vegetative phase.

2. Seedling Culture

Both hydroponics and substrate culture are suitable for the soilless culture of lettuce. For hydroponics, flat-bottomed impermeable plastic trays that are 60 cm long, 24 cm wide, and 4 cm high can be selected as seedling trays, and square sponge blocks (3 cm^2) can be used as the substrate for fixing seeds. Two to three seeds are sown on each sponge block, and then water is poured into the seedling trays until the surface of the sponge blocks is immersed. To increase humidity, the seeds should be sprayed twice to three times a day until they germinate. For substrate culture, the substrate may be made by mixing the peat and vermiculite at a ratio of 1 : 1.

3. Planting

Seedlings with 3–4 true leaves are appropriate for planting. If sponge blocks are used, the sponge blocks together with seedlings can be directly placed in the planting cups. If substrate culture is selected, seedlings should be removed from the plug trays to wash off the substrate from their root systems with clean water and then placed in the planting cups by fixing them with sponge blocks.

4. Management after Planting

Since the root systems of the seedlings are short and small in the early stage of planting, the water level should be high, preferably 0.5–1 cm from the cover plate. Otherwise, the root systems cannot absorb nutrients and water, which will eventually lead to withering and death of the seedlings. As the root system grows, the liquid level can be gradually lowered to increase the distance between the liquid level and the cover plate. An appropriate formulation should be selected to prepare a balanced nutrient solution for the lettuce. According to the climate and lettuce growth conditions, the consumption of the nutrient solution should be measured and recorded regularly to test the conductivity and pH level. The optimal EC values for hydroponic lettuce are 1.6–1.8 mS/cm in winter and 1.4–1.6 mS/cm in summer, and the optimal pH level is 6–7. The nutrient solution is supplied through circulation, and a timer is used to control the supply duration, i.e. 30 minutes every 2 hours from 8:00 to 15:00 every day. The nutrient solution should be replaced once after consecutive cultivation for 3–4 times. After the previous crop of lettuces is harvested, residual roots and other debris in the growing channels should be removed, and water and nutrient solution should be supplemented before the next crop can be cultivated.

5. Harvest

Non-head lettuce can be harvested according to market demands. During production, phased sowing and harvesting in batches may be adopted to realize an annual balanced supply.

Task 4 Organic Culture of Lettuce

I. Concept of Organic Production

Organic production is an agricultural production mode in which specific production principles together with natural laws and ecological principles are followed to maintain a stable production system and coordinate planting and animal husbandry without the use of genetically modified organisms or their products, or chemically synthesized pesticides, chemical fertilizers, growth regulators, feed additives and the like.

II. Concept of Organic Lettuce

Organic lettuce, cultivated under organic production conditions, is an annual or biennial herbaceous plant in the genus *Lactuca*, family Asteraceae, with shallow and dense roots, mostly distributed in a soil layer of 20–30 cm thick. Organic lettuce prefers a cool climate. Short days of lighting and low temperature are conducive to its growth and development, while long days of lighting and high temperature will easily make it bolt and flower. Organic lettuce is crisp and tender, with a high water content. A uniform and sufficient supply of water is required throughout its vegetative phase. Specific varieties include organic asparagus lettuce, organic lettuce, and organic iceberg lettuce.

III. Essentials for Organic Culture of Lettuce

1. Cultivation Season

Organic lettuce has wide adaptability. According to the market demand, different varieties can be selected for planting all year round to realize an annual balanced supply. Leaf lettuce has good adaptability and is mainly sown in autumn and winter and harvested in winter and spring. Head lettuce has a narrow temperature tolerance range and is neither resistant to low temperature nor high temperature. In

North China, it is mainly cultivated in spring in the open field, but can be cultivated throughout the year in protected fields. In South China, it is sown in autumn and winter and harvested in spring or sown in autumn and harvested in winter. In South China, sowing is implemented from August to the following February and harvested from September to the following April. In summer, shading and rain proof facilities are used for cultivation.

2. Land Preparation and Fertilization

The soil for cultivation should have a loose texture, contain abundant organic matter, and drain well with strong water and nutrient retention capacity. The land should be deeply plowed, solarized, and then applied with decomposed farmyard manure, commercial organic fertilizer, or rapeseed cakes. Further land preparation is required 7–10 days after that by making borders with a width of 1.8–1.9 m. Low borders should be made for regular-season cultivation, and high borders for off-season cultivation. It is necessary to adopt the strategy of applying sufficient base fertilizer and supplementing fertilizer as appropriate.

3. Sowing

When the seedling land is ready, direct-sowing cultivation is mostly adopted. Small arched sheds with plastic film mulching are used for thermal insulation in early spring, autumn, and winter. Small arched sheds with shade nets are used in summer. An area of 15–20 m^2 of seedling land and 25–30 g seeds are needed for cultivation in each mu of field. Generally, transplanting is performed at a planting spacing of 32 cm× 35 cm, with 4,800–5,200 plants/mu planted. Seedlings with enlarged stems should be discarded during transplanting to avoid early bolting.

4. Field Management

Timely water the seedlings, control the water and nutrient content, and inter-till the land frequently after transplanting. Water the seedlings again and apply supplemental fertilizer from the rosette stage to when the leaves seal the ridges, and afterward water them once the surface soil becomes dry. When the stem begins to enlarge, it is necessary to apply the supplemental fertilizer the third time.

5. Disease and Pest Control

(1) Main diseases and pests.

Common diseases include virus disease, soft rot, ripe rot, downy mildew, and *Rhizoctonia* rot. The main insect pests include cabbage moths (*Plutella xylostella*), *Phyllotreta striolata*, aphids, cotton leafworms, beet armyworms, and cabbage caterpillars.

(2) Prevention and control strategies.

Virus diseases are mainly prevented and controlled by cultivating disease-resistant varieties with measures to avoid aphid damage.

For the prevention and control of soft rot and other viruses, certified organic biological agents should be used in combination with good cultivation techniques.

For the prevention and control of main pests such as cabbage moths and striped flea beetles, it is necessary to master their occurrence regularity and resistance to pesticides and apply organic biological agents together with cultivation techniques such as combining different varieties for cultivation.

(3) Prevention and control methods.

Select disease-resistant varieties according to local conditions.

①Cultivation prevention and control: Avoid continuous cropping and implement crop rotation. After harvesting the previous crop of vegetables, timely plow and sun the borders, remove plant residues and old leaves, keep the field clean, and reduce the residual germs, insects, and eggs.

②Biological prevention and control: Prevent and control the pests with biological agents such as matrine, rotenone, natural pyrethrins, and B.t.

③Physical prevention and control: Prevent and control the pests with fly nets, yellow sticky boards, and solar pest-killing lamps.

6. Harvest

The organic lettuce can be harvested when its main stem is as long as its longest leaf. At this time, the long tender stems taste crispy with good quality and can be marketed once the old leaves are removed.

[Summary]

I. Key Points

(1) Master the advantages, disadvantages, and characteristics of the DFT;

(2) Master the cultivation management of the DFT;

(3) Master the advantages, disadvantages, and characteristics of the NFT.

II. Difficult Points

(1) Master the management of timers, conductivity meters, pH meters, temperature meters, and other devices used for cultivation with the nutrient film technique;

(2) Master the management of aeroponics nutrient solutions.

[Skill Training]

Skill Training 5-1: Preparation of Nutrient Solution

I. Purposes and Requirements

(1) Master the preparation methods of mother liquor and working solution;

(2) Understand the principles for preparing nutrient solutions;

(3) Be skilled in the operation and operate according to relevant procedures and standards;

(4) Prepare the nutrient solutions accurately without precipitation;

(5) Clearly identify the nutrient solutions and keep complete records.

II. Planning

1. Materials and Reagents

Reagents or fertilizers required for preparing the mother liquor of the Japanese garden test formula (Table 5-1): 1 mol/L NaOH and 1 mol/L HNO_3 solution.

Table 5-1 Formula of Japanese Garden Test Nutrient Solution

Name of Salt Compound	Amount Used (mg/L)
$Ca(NO_3)_2 \cdot 4H_2O$	945.00
KNO_3	809.00
$NH_4H_2PO_4$	153.00
$MgSO_4 \cdot 7H_2O$	493.00
EDTA-Na_2Fe	20.00
H_3BO_3	2.86
$MnSO_4 \cdot 4H_2O$	2.13
$ZnSO_4 \cdot 7H_2O$	0.22
$CuSO_4 \cdot 5H_2O$	0.08
$(NH_4)_4Mo_7O_{24} \cdot 4H_2O$	0.02

2. Instruments and Tools

Tray balances or platform scales, electronic analytical balances (sensitivity 0.001 g), water pumps, acidimeters, conductivity meters; magnetic stirrers, black plastic buckets (50 L, 2 pieces), plastic beakers (500 mL, 1,000 mL) or plastic basins, black plastic liquid storage tanks (50L, 3 pieces), plastic water pipes, label paper, glass rods, short wooden sticks or plastic rods, pens, marker pens, record forms for preparation of mother liquor, and record forms for preparation of working solution.

III. Procedures

1. Preparation of Mother Liquor (Fig. 5-4)

(1) Calculation: Determine the type, concentration ratio, and amount of nutrient solution formula and mother liquor to be prepared and then calculate the amount of reagent or fertilizer needed. The following mother liquors are required to be prepared according to the requirements for the garden test formula: 10 L and 100-fold-concentrated mother liquor A [$Ca(NO_3)_2 \cdot 4H_2O$ and KNO_3], mother liquor B ($NH_4H_2PO_4$ and $MgSO_4 \cdot 7H_2O$), and 1 L and 1,000-fold-concentrated

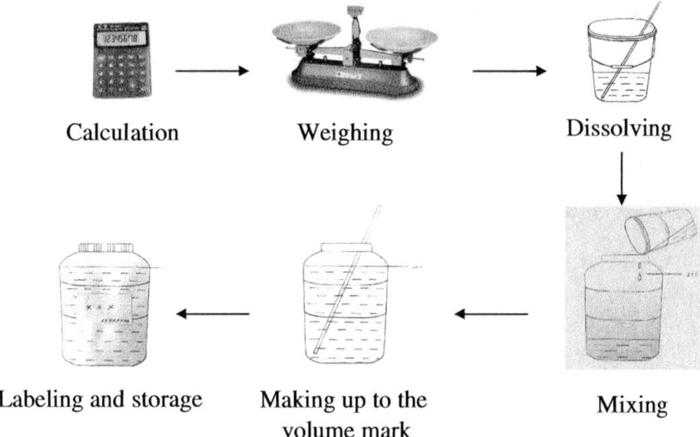

Fig. 5-4 Preparation Process of Mother Liquor

mother liquor C (EDTA-Na_2Fe and various trace element compounds). The calculated amount of reagent or fertilizer is as follows: $Ca(NO_3)_2 \cdot 4H_2O$, 945.00 g; KNO_3, 809.00 g; $NH_4H_2PO_4$, 153.00g; $MgSO_4 \cdot 7H_2O$, 493.00g; EDTA-Na_2Fe, 20.00g; $MnSO_4 \cdot 4H_2O$, 2.13 g; H_3BO_3, 2.86g; $ZnSO_4$,$\cdot 7H_2O$, 0.22g; $CuSO_4 \cdot 5H_2O$, 0.08g; $(NH_4)_4Mo_7O_{24} \cdot 4H_2O$, 0.02 g.

(2) Weighing: Weigh reagents or fertilizers with the platform scale, tray balance, or analytical balance, and place them in such clean containers as beakers and plastic basins. Ensure stable, accurate (to within ±0.1 g), and fast operation when weighing.

(3) Dissolving and mixing: Prepare three types of mother liquors A, B, and C in liquid storage tanks A, B, and C, respectively. Prepare mother liquor A in reservoir A by dissolving together the reagents or fertilizers that do not precipitate with calcium salts. Prepare mother liquor B in reservoir B by dissolving together the reagents or fertilizers that do not precipitate with phosphate. Prepare mother liquor C in reservoir C by dissolving chelated iron and other trace element compounds, respectively. If there is no readily available chelated iron reagent, $FeSO_4 \cdot 7H_2O$ and EDTA-Na_2 can be used. The preparation method is: To prepare mother liquor C, weigh 13.9 g of $FeSO_4 \cdot 7H_2O$ and 18.6 g of EDTA-Na_2, respectively, dissolve them with warm water, then slowly pour the $FeSO_4 \cdot 7H_2O$ solution into the EDTA-Na_2 solution, stir while adding to mix them well, then pour the mixture into reservoir C, then slowly pour

the various trace element compound solutions dissolved respectively into reservoir C, stir while adding, and finally add water to the final volume to obtain the 1,000-fold mother liquor C; To prepare mother liquor A, first dissolve Ca $(NO_3)_2 \cdot 4H_2O$ and KNO_3, respectively, and then mix them together; To prepare mother liquor B, first dissolve $NH_4H_2PO_4$ and $MgSO_4 \cdot 7H_2O$, respectively, and then mix them together. The mother liquor C can be used in the soilless cultivation of any crop.

(4) Making up to the volume mark: Respectively inject clear water into the liquid storage tanks A, B, and C to the volume required to be prepared and mixed well.

(5) Labeling and storage: Affix labels or use a marking pen to indicate the name of mother liquor, mother liquor number, concentration ratio or concentration, preparation date, and operator on the liquid storage tanks A, B, and C, and then store them in a cool and dark place. If the mother liquor is stored for a long time, it should be acidified to prevent precipitation. Generally, it can be acidified to pH 3–4 with HNO_3.

(6) Recording: After the preparation of mother liquor, carefully fill in the record form for preparation of mother liquor. See Table 5-2 for the form format.

Table 5-2 Record Form for Preparation of Mother Liquor

Formula name			Used by	
Mother liquor A	Concentration ratio		Preparation date	
	Volume		Calculated by	
Mother liquor B	Concentration ratio		Reviewed by	
	Volume		Prepared by	
Mother liquor C	Concentration ratio		Remarks	
	Volume			
Name and weighed amount of raw materials				

2. Preparation of the Working Solution

(1) Dilution of concentrate: This is a commonly used method of preparing working solutions in production. See Fig. 5-5 for the preparation method.

① Calculate the pipette volume of all three types of mother liquors. The formula for using the following volume is as below:

$$V_2 \text{ (Pipette volume of mother liquor)} = \frac{V_1 \text{ (Volume of working solution)}}{n \text{ (Concentration ratio of mother liquor)}}$$

For this training, 100 L of working solution is prepared with the above mother liquor. According to the calculation results, 1 L of mother liquors A and B, respectively, and 0.1 L of mother liquor C should be pipetted.

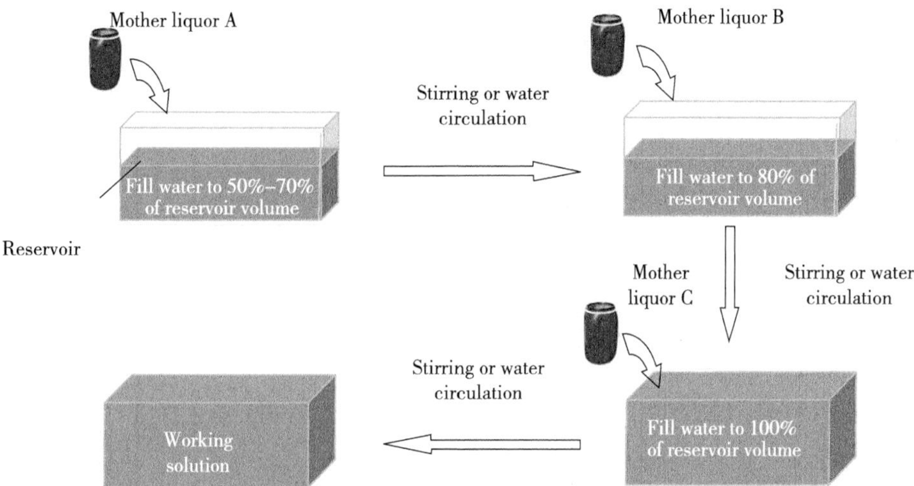

Fig. 5-5 Operating Procedure for Preparation of the Working Solution

② Fill water to 50%–70% of reservoir volume (volume of the nutrient solution to be prepared).

③ Pipette mother liquor A and inject it into the reservoir. Start the water pump to circulate the nutrient solution in there for 30 min or stir to allow it to spread evenly.

④ Pipette mother liquor B, and slowly inject it into the reservoir from the clear water inlet of the reservoir to dilute it by the running flow before it spreads in the reservoir; start the water pump to circulate the nutrient solution in the reservoir for 30 min or stir to allow it to spread evenly. The amount of water added in this process reaches 80% of the solution volume.

⑤ Pipette mother liquor C and add it into the reservoir in the same way as adding mother liquor B. Circulate the solution through the water pump or stir it well. The water volume reaches 100% in this case.

⑥ Measure the pH and EC values of the nutrient solution using an acidimeter and a conductivity meter. If the measured pH value fails to meet the requirements for formula and crop cultivation, adjust it timely. After the pH value is adjusted for the nutrient solution, stand it for more than half an hour before use. Then, circulate it on the planting bed for about 5–10 min, and test the pH value again until it meets the requirements.

⑦ Fill in the record form for preparation of the working solution for inspection. See Table 5-3 for the record form for the preparation of the working solution.

Table 5-3 Record Form for Preparation of Working Solution

Formula name		Used by	
Nutrient solution volume		Preparation date	
Calculated by		Reviewed by	
Prepared by		Water pH value	
EC value of the nutrient solution		pH value of Nutrient solution	
Name and measured (pipetted) amount of raw materials			

(2) Direct preparation: In production, if a large amount of working solution is required, it may be prepared directly after weighing raw materials containing macroelements or prepared by diluting mother liquor C prepared in advance if the raw materials contain micronutrients. In this training, the 100 L of garden test formula nutrient solution can also be prepared directly. The specific procedures are as follows.

① Calculate the amount of reagent or fertilizer required according to the nutrient solution formula and the volume of nutrient solution to be prepared.

② Fill the reservoir with 50%–70% water.

③ Weigh corresponding amounts of reagents or fertilizers for the preparation of mother liquor A, dissolve them in the plastic basin, and then pour them into the reservoir. Start the water pump to circulate the nutrient solution for 30 min or stir it well.

④ Weigh corresponding amounts of compounds for the preparation of mother liquor B, dissolve them in a plastic basin, and pour them into the reservoir from near the water inlet to allow them to be diluted by the running water before they spread in the reservoir; start the water pump to circulate the nutrient solution for 30 min or stir it well. The amount of water added in this process reaches 80% of the total solution volume.

⑤ Measure and dilute the pre-prepared mother liquor C. Then, pour it slowly from the water inlet of the reservoir and start the water pump to circulate the nutrient solution for 30 min or until the solution is well mixed.

⑥ and ⑦ are the same as the ⑥ and ⑦ of the dilution of the concentrate method.

IV. Precautions

(1) Check the calculated amount of reagents or fertilizers repeatedly to ensure accuracy. Ensure the accuracy of weighing and consistency between descriptions and real items.

(2) The following matters should be paid attention to in the calculation of the amount of reagents or fertilizers: ①The fertilizers used for soilless culture are mostly for agricultural and industrial uses and often contain hygroscopic water and other impurities, with low purity. The amount should be corrected according to the actual purity. ②Ca^{2+} and Mg^{2+} contained in the water should be deducted in hard water areas. For example, Ca^{2+} and Mg^{2+} in the formula are provided by $Ca(NO_3)_2 \cdot 4H_2O$ and $MgSO_4 \cdot 7H_2O$, respectively, and the actual amount of $Ca(NO_3)_2 \cdot 4H_2O$ and $MgSO_4 \cdot 7H_2O$ should be the formula amount minus the amount of Ca^{2+} and Mg^{2+} contained in the water. However, the amount of nitrogen in $Ca(NO_3)_2 \cdot 4H_2O$ is reduced after the deduction of Ca^{2+}, and the reduced amount can be supplemented with nitric acid (HNO_3), which not only plays a role in supplementing the nitrogen but also neutralizes the alkalinity of the hard water. If the pH value of water is still not lowered to the ideal level after nitric acid is

added, the amount of phosphate can be appropriately reduced, and phosphoric acid can be used to neutralize the alkalinity of the hard water. If the nutrient solution is too acidic, the amount of potassium nitrate can be increased to supplement nitrate nitrogen and reduce the amount of potassium sulfate accordingly. Deducting the amount of magnesium in nutrition, the actual amount of $MgSO_4 \cdot 7H_2O$ is reduced, and so is the amount of SO_4^{2-}. However, since hard water itself contains a large amount of sulfate, additional supplementation is generally not required. If required, a small amount of H_2SO_4 can be added. In hard water areas, the amount of calcium nitrate is small, and the amount of phosphorus and nitrogen lacking is supplemented by supplying nitric acid and phosphoric acid.

(3) Supplies, including weighed fertilizers used for preparation of the nutrient solution, should be placed on the preparation site in an orderly manner, and preparation may not begin before it is verified that there is nothing missing. Do not rush to operate when materials are not ready.

(4) The container used for dissolving reagents or fertilizers should be swabbed with clean water, and the water used should be poured into the liquid storage tank or reservoir.

(5) To accelerate the dissolution of reagents or fertilizers, warm water can be used for dissolution, or a magnetic stirrer can be used for stirring.

(6) Upon preparation of the working solution, avoid adding the mother liquor too quickly, which may cause many precipitates due to high local concentration. If the precipitates cannot be dissolved after pump circulation for a long time, the nutrient solution should be discarded and prepared again.

(7) Establish record files for inspection.

V. Answer Questions

(1) What is the purpose of mother liquor preparation?

(2) What are the consequences of inaccurate nutrient solution preparation?

(3) What are the differences between the nutrient solution used for soilless culture and the medium used for tissue culture?

Skill Training 5-2: Effects of Photoperiod on Lettuce Seedling Growth

I. Purposes and Requirements

(1) Master the regulation of lettuce growth photoperiod;

(2) Be able to skillfully determine the morphology index, yield index, quality properties, etc. of lettuce;

(3) Be able to analyze differences in lettuce growth according to different photoperiods.

II. Planning

1. Materials and Reagents

The looseleaf lettuces Hong Kong Iceberg and Grand Rapids will be provided by the Beijing Academy of Agriculture and Forestry Sciences. Sponge blocks 3 cm thick will be used for the nursery. The formula for cultivating Japanese Yamazaki Lettuce (lettuce): calcium nitrate tetrahydrate [$Ca(NO_3)_2 \cdot 4H_2O$] 236 mg/L, potassium nitrate (KNO_3)404 mg/L, ammonium dihydrogen phosphate ($NH_4H_2PO_4$) 57 mg/L, and magnesium sulfate heptahydrate ($MgSO_4 \cdot 7H_2O$) 123 mg/L. The formula of general trace elements: sodium feredetate (EDTA-Na_2Fe) 20–40 mg/L, boric acid 2.86 mg/L, manganese sulfate tetrahydrate ($MnSO_4 \cdot 4H_2O$) 2.13 mg/L, zinc sulfate heptahydrate ($ZnSO_4 \cdot 7H_2O$) 0.22 mg/L, copper sulfate pentahydrate ($CuSO_4 \cdot 5H_2O$) 0.08 mg/L, and ammonium molybdate tetrahydrate [$(NH_4)_4Mo_7O_{24} \cdot 4H_2O$] 0.02 mg/L. The nutrient solution will be prepared using analytical chemical reagents.

T4 fluorescent lamps with a power of 20 W will be used as the light source in the test, and 20 lamps will be set per plot, with a photon flux density of 80 $\mu mol \cdot m^{-2} \cdot s^{-1}$.

2. Test Method

Cut the prepared 3-cm-thick loose sponge block into small 3-cm-thick blocks, and keep them connected during cutting to facilitate stacking. After the sponge blocks are cleaned, stack them flatly in the water-tight seedling tray. Make a 1cm deep hole in the center of each sponge block and sow three seeds in each hole; thin the seedlings to 1 plant after the seeds germinate and 3 leaves emerge; then transplant the lettuce when it grows to have five unfolded leaves. The hydroponic nutrient solution tank is 180 cm long, 60 cm wide, and 12 cm high. On the tank, there is a PVC cultivation plate with 28 holes, each with a diameter of 32 mm. Add 100 L of nutrient solution in each cultivation tank, and cultivate 28 lettuce plants per cultivation plate. Each cultivation tank is considered as one plot.

Set three photoperiods (12 h, 16 h, and 20 h) for the 2 test varieties, i.e. 6 treatments in total and 3 repeats for each treatment. Control the indoor environment by air conditioning, control the daytime temperature at 24°C and night temperature at 18°C, and control the daytime relative humidity at 48%–52% and night relative humidity at 60%–70%. Supplement nutrient solution once a week, ventilate for 1 h in the three periods: 7:00–8:00, 11:00–12:00, and 17:00–18:00 every day, control the EC value within 1.3–1.8 ms/cm and adjust the pH value to 6.0–6.8 with phosphoric acid.

3. Index Measurement

When lettuce is harvested, randomly select six lettuce plants to measure the plant height, the number of leaves, length and width of the seventh leaf, fresh weight of the aboveground and underground parts, dry weight of the aboveground part, dry weight of the underground part, root shoot ratio and other indexes. The measurement method for the dry weight of the aboveground part: Expose the aboveground part of the lettuce sample under 105°C for 1 h to kill it, and dry it to a constant weight at 70°C. The calculation formula for the moisture content of the aboveground part is as follows.

$$\theta = \frac{G}{W} \times 100\%$$

In this formula, θ is the moisture content of the aboveground part; G is the dry weight of the aboveground part; W is the fresh weight of the aboveground part.

The calculation formula for root shoot ratio is:

$$\Phi=\frac{S}{W}$$

Where Φ is the root shoot ratio; S is the fresh weight of the underground part; W is the fresh weight of the aboveground part.

III. Implementation

Perform lettuce hydroponic culture in groups and carry out routine management of lettuce as described above, measure the above indexes after harvest, and record the data in Table 5-4.

Table 5-4 Measurement of Lettuce Indexes for Different Photoperiods

Index	12 h	16 h	20 h
Plant height			
Number of leaves			
Length of the seventh leaf			
Width of the seventh leaf			
Fresh weight of the aboveground part			
Fresh weight of the underground part			
Dry weight of the aboveground part			
Dry weight of the underground part			
Root shoot ratio			

Skill Training 5-3: Effects of Plant Growth Regulators on Lettuce Seedling Growth

I. Purposes and Requirements

(1) Understand the roles of plant growth regulators in plant production;

(2) Master the preparation method of plant growth regulators;

(3) Learn to use plant growth regulators correctly.

II. Planning

1. Materials

Lettuce seeds and plant growth regulators IBA, IAA, and 6-BA.

2. Treatment of Lettuce

At first, soak the lettuce seeds at room temperature for 1 h, pick out full and shiny ones and dry their surface with absorbent paper. Then, soak them in 1/5,000 potassium permanganate solution for 1 h for sterilization and take them out and wash them with distilled water. At last, dry the seed surface again with absorbent paper.

3. Test Design

Establish five concentration levels (10, 15, 20, 25, and 30 mg/L) for plant growth regulators IBA, IAA, and 6-BA, respectively. Soak some seeds in the plant growth regulators with corresponding concentration levels respectively for 6 h, and soak some seeds in distilled water for 6 h as the control group. After soaking, rinse the seeds with distilled water and transfer them into each Petri dish padded with gauze and filter paper, respectively, with 30 seeds in each Petri dish. Culture 5 dishes for each treatment, and culture in a dark, 20°C incubator. Care should be taken to supplement water and keep the gauze and filter paper wet during culture.

4. Measurement of Growth Indexes

Observe and record seed germination every afternoon, calculate the germination rate on the fourth day, measure seedling root length, fresh weight of root, cotyledon weight, and fresh weight of young plant when seedlings have 5–6 leaves, and calculate the germination index, viability index (VI), and simplified viability index. The calculation formula is:

$$GI = \sum (Gt/Dt)$$

In the formula, Gt is the number of germinations in time t, Dt is the corresponding days of germination.

$$VI = S \times \sum (Gt/Dt)$$

S is the length of the seedling radicle.

$$\text{Simplified viability index} = G \times S$$

G is the germination rate, S is the length of the seedling radicle.

[Extension Tasks]

I. Review Exercises

(1) What are the differences between the deep flow technique and the nutrient film technique?

(2) What are the characteristics and advantages of aeroponics?

(3) What are the characteristics of cropping patterns in open-field lettuce cultivation in southern and northern China?

(4) How can the nutrient solution be supplied intermittently during cultivation using the nutrient film technique?

(5) What are the characteristics of rock wool? What are the differences between open-loop and closed-loop rock wool culture?

(6) Explore the specific application of several soilless culture methods in production.

II. More to Know

(I) Common Fertilizers for Soilless Culture

1. Nitrogen Source

There are two sources, namely, nitrate nitrogen and ammonium nitrogen. Most vegetables are crops favoring nitrate nitrogen. However, Although excessive nitrate nitrogen will not result in harm, excessive ammonium nitrogen will hinder the growth. Common nitrogen source fertilizers include calcium nitrate, potassium nitrate, ammonium dihydrogen phosphate, ammonium sulfate, ammonium chloride, and ammonium nitrate.

(1) Calcium nitrate [$Ca(NO_3)_2 \cdot 4H_2O$].

It contains two nutrient elements, nitrogen and calcium, of which nitrogen (N) content is 11.9% and calcium (Ca) content is 17.0%. Calcium nitrate is a white crystal and is readily soluble in water: 129.3 g of calcium nitrate can be dissolved per 100 mL of water at 20°C. It is extremely hygroscopic and deliquesces easily, especially when exposed to the air or placed under high temperatures and high humidity. Therefore, it should be stored in a sealed and cool place.

Calcium nitrate is a physiological alkaline salt. This is because the crop root system uptakes the nitrate ions more quickly than calcium ions. Nevertheless, the physiological alkalinity is not very high since calcium ions are also absorbed by crops and will gradually drop as more calcium ions are absorbed. Calcium nitrate is the most widely used in preparing nitrogen and calcium fertilizers in the soilless culture. This is especially true for calcium fertilizers (nutrient solutions), most of which are prepared with calcium nitrate.

(2) Ammonium nitrate [NH_4NO_3].

The content of nitrogen in ammonium nitrate is 34%–35%, of which ammonium nitrogen (NH_4^+ -N) and nitrate nitrogen (NO_3^- -N) account for half each. Ammonium nitrate is a white crystal. For agricultural and industrial uses, ammonium nitrate is often made into particles by adding hydrophobic substances. It has a high dissolubility: 188 g of ammonium nitrate can be dissolved per 100 mL of water at 20°C.

Ammonium nitrate is highly hygroscopic and easy to harden. Pure ammonium nitrate is likely to absorb moisture and deliquesce when exposed to air. Therefore, it should be stored in a sealed and cool place. In addition, ammonium nitrate, being flammable and explosive, must not be stored or transported with other flammable and explosive substances. The ammonium nitrate caked when moisturized should not come under violent strikes by metal objects such as steel hammers but can be smashed gently with a wooden or rubber hammer or other non-metallic materials.

Ammonium nitrate contains 50% ammonium nitrogen and 50% nitrate nitrogen. As most crops absorb ammonium ions more quickly than nitrate ions early when ammonium nitrate is applied, they show a high physiological acidity, which, however, gradually disappears when they absorb an equivalent amount of ammonium nitrogen and nitrate nitrogen. In addition, attention should be paid to the amount of ammonium nitrate applied as a source of nitrogen for nutrient solutions since a high dosage may affect the nutrient uptake and growth of crops sensitive to ammonium nitrogen.

(3) Potassium nitrate [KNO_3].

Potassium nitrate, with a nitrogen (N) content of 13.9% and a potassium (K) content of 38.7%, can serve as a source of nitrogen and potassium. It appears as white crystals with low hygroscopicity and will cake during long-term storage in a humid environment. It is freely soluble in water: 31.6 g can be dissolved per 100 mL of water at 20°C. Potassium nitrate is flammable and explosive, so it should be noted that violent impacts must be avoided during storage and transportation. It must not be mixed with other flammable and explosive materials. Potassium nitrate is a physiologically alkaline fertilizer.

(4) Ammonium sulfate [$(NH_4)_2SO_4$].

Ammonium sulfate, with a nitrogen (N) content of 20%–21%, is made by neutralizing NH_3 with sulfuric acid. It appears as white crystals and is freely soluble in water: 75 g can be dissolved per 100 g of water at 20°C. Ammonium sulfate shows excellent physical properties and poor hygroscopicity. However, when ammonium sulfate contains rich free acid or is stored under high air humidity for a long time, it absorbs moisture and cake.

During the absorption of ammonium sulfate by plants, rich sulfuric acid will be accumulated to make the solution acidic as the root system of most crops absorbs NH^{4+} at a faster rate than SO_4^{2-}. Therefore, ammonium sulfate is considered a physiologically acidic fertilizer. Attention should be paid to the changes in its physiological acidity when it is used as a nitrogen source for nutrient solutions.

2. Phosphorus Source

Commonly used phosphate fertilizers include ammonium dihydrogen phosphate, diammonium phosphate, potassium dihydrogen phosphate, and calcium superphosphate. Excessive phosphorus may lead to iron and magnesium deficiency.

(1) Ammonium dihydrogen phosphate [$NH_4H_2PO_4$].

Also known as monoammonium phosphate or MAP, it is made by introducing ammonia gas into phosphoric acid. Pure ammonium dihydrogen phosphate appears as white crystals, while fertilizer ammonium dihydrogen phosphate is mostly gray crystals. The pure product contains 61.7% phosphorus (P_2O_5) and 11%–13% nitrogen (N). It is freely soluble in water (high dissolubility): 36.8 g can be dissolved per 100 mL of water at 20°C. It provides both nitrogen and phosphorus as nutrient elements and buffers pH changes in the solution.

(2) Diammonium hydrogen phosphate [$(NH_4)_2HPO_4$].

Also known as diammonium phosphate or DMP, it is made by introducing ammonia gas into a phosphoric acid solution. Pure diammonium hydrogen phosphate appears as white crystals. The pure product contains 53.7% phosphorus (P_2O_5) and 21% nitrogen (N). Fertilizer diammonium hydrogen phosphate often contains a certain amount of ammonium dihydrogen phosphate, with a phosphorus (P_2O_5) content of 20% and nitrogen (N) content of 18%. It can buffer pH changes in nutrient solution or substrate.

(3) Potassium dihydrogen phosphate [KH_2PO_4].

It appears as white crystals or powder, with a molecular weight of 136.09 and is freely soluble in water: 22.6 g can be dissolved per 100 mL of water at 20°C. Potassium dihydrogen phosphate is stable and robust, and not prone to deliquescence, but it may also absorb moisture and cake when stored in places with high humidity. Since dissociated phosphate radicles have different valencies when potassium dihydrogen phosphate dissolves in water, it can buffer pH changes in solution to some extent and provide both potassium and phosphorus nutrient elements, making it an important phosphorus source in soilless culture.

(4) Calcium superphosphate [$Ca(H_2PO_4)_2 \cdot H_2O+CaSO_4 \cdot H_2O$].

Calcium superphosphate is also known as single superphosphate or SSP. It is made by adding sulphuric acid to the crushed phosphate powder for dissolution and contains monocalcium phosphate [$Ca(H_2PO_4)_2$], an active component containing phosphorus, and calcium sulfate (gypsum, $CaSO_4 \cdot H_2O$) generated during manufacturing, both of which account for 30%–50% and 40% of the fertilizer weight, respectively. The rest are impurities. Calcium superphosphate appears as gray or ash black granules or powder. First-grade calcium superphosphate contains 18% available phosphorus (P_2O_5), <4% free acid, <10% moisture, 19%–22% Ca, and 10%–12% S. Calcium superphosphate is a water-soluble phosphate fertilizer. When it is dissolved in water, some precipitates will remain at the bottom of the container. These precipitates are insoluble calcium sulfate, so do not mistake calcium superphosphate as a slow-acting or insoluble fertilizer.

Fe-based, Al-based, and other compounds originally existing in the original phosphate ore are also found in the fertilizer after the ore is dissolved with sulfuric acid during the manufacturing of calcium superphosphate. When calcium superphosphate absorbs moisture, monocalcium phosphate will react with such compounds and form insoluble iron phosphate, aluminum phosphate, and other compounds, thus reducing the effect of phosphoric acid, which is called the degradation of phosphoric acid. Therefore, calcium superphosphate shall be stored somewhere dry to prevent moisture absorption and reduce the fertilization efficiency of calcium superphosphate.

In the soilless culture, calcium superphosphate is mainly used as a source of phosphorus and calcium after being mixed in the substrate during substrate cultivation and seedling cultivation. Since it contains rich free sulfuric acid and other impurities as well as calcium sulfate precipitates, it is generally not used as a fertilizer source for preparing nutrient solutions in hydroponics.

3. Potash Fertilizer

Commonly used potash fertilizers include potassium nitrate, potassium sulfate, potassium chloride, and potassium dihydrogen phosphate. Plants absorb potassium quickly, so timely replenishment shall be made. However, excessive potassium ions in the nutrient solution may affect the absorption of calcium, magnesium, and manganese.

(1) Potassium nitrate [KNO_3]: See the "Nitrogen Source" section above.

(2) Potassium sulfate [K_2SO_4].

Pure potassium sulfate appears as white powder or crystals, while the potassium sulfate used as agricultural fertilizer is mostly white or light yellow powder. Pure potassium sulfate contains 54.1% potassium (K_2O). Fertilizer potassium sulfate contains 50%–52% potassium (K_2O) and 18% sulfur (S). It is soluble in water (slightly low dissolubility): 11.1 g can be dissolved per 100 mL of water at 20°C. It shows low hygroscopicity and does not cake with good physical properties. As a physiologically acidic fertilizer, it forms a neutral solution when dissolved in water.

(3) Potassium chloride [KCl].

Pure potassium chloride appears as white crystals, while fertilizer potassium chloride is often purplish-red, light yellow, or white powder, which is attributed to the color of the minerals of different sources at the time of production. Potassium chloride contains 50%–60% potassium (K_2O) and 47% chlorine. It is freely soluble in water: 34.4 g can be dissolved per 100 mL of water at 20°C. As a physiologically acidic fertilizer with low hygroscopicity, it forms a neutral solution when dissolved in water. Potassium chloride can also be used as a source of potassium in soilless culture but is seldom applied as it contains rich chloride ions (Cl^-), which has a negative impact on the yield and quality of "chlorine-sensitive crops" such as potatoes and beets.

(4) Potassium dihydrogen phosphate: See the "Phosphorus Source" section above.

4. Calcium Source

The most commonly used calcium fertilizer is calcium nitrate. Calcium chloride and calcium superphosphate can also be used as appropriate. Calcium hardly moves within the plant, which explains the frequent occurrence of calcium deficiency symptoms during soilless culture. Therefore, special attention should be paid to taking adjustment measures.

(1) Calcium nitrate: See the "Nitrogen Source" section above.

(2) Calcium chloride [$CaCl_2$].

Containing 36% calcium (Ca) and 64% chlorine (Cl), it appears as white powder or crystals. With high hygroscopicity, it has a high solubility in water and forms a neutral aqueous solution. As a physiologically acidic fertilizer, it is rarely used as a calcium source in soilless culture but is mostly used as foliar fertilizer when the crop suffers calcium deficiency. It is also applied in formulas where calcium nitrate is not used as the calcium source. It should not be used on "chlorine-sensitive crops" and should be used with caution on other crops.

(3) Calcium sulfate [$CaSO_4 \cdot 2H_2O$].

Calcium sulfate, also known as gypsum, appears as a white powder and contains 23.28% calcium (Ca) and 18.62% sulfur (S). It is made of crushed or heated gypsum ore. There are three types of gypsum for agricultural use, namely dihydrate gypsum ($CaSO_4 \cdot 2H_2O$), calcined gypsum ($CaSO_4 \cdot 1/2H_2O$), and phosphogypsum ($CaSO_4 \cdot 2H_2O$). The solubility of calcium sulfate is extremely low, with only 0.204 g dissolved per 100 mL of water at 20°C. It forms a neutral aqueous solution and is a physiologically acidic fertilizer. Calcium sulfate is rarely used in the preparation of nutrient solutions in hydroponics and may be used as calcium salt in a few formulas. It is generally mixed into the substrate in substrate cultivation as a supplement to the calcium source.

5. Iron Source

Iron deficiency can be caused by high pH levels, insufficient potassium as well as excessive phosphorus, copper, zinc, and manganese in the nutrient solution. To ensure iron supply, chelated iron is generally used. Chelated iron, a chelate formed by organic compounds and trace element iron, tends not to be fixed or precipitated in the nutrient solution and is easily absorbed by crops. Its effect is markedly better than inorganic iron salts and organic iron acids. Commonly used chelated irons include sodium feredetate (EDTA-Na_2Fe). Sodium feredetate [EDTA-Na_2Fe], with a molecular weight of 390, contains 14.32% iron, appears as an earthy yellow powder, and is freely soluble in water. Ethylene diamine tetraacetic acid and monosodium ferric salt [EDTA-NaFe] are also used occasionally. The amount of chelated iron is generally 3–5 mg per liter of the nutrient solution when calculated by the weight of iron.

6. Magnesium, Zinc, Copper, Manganese, and Other Sulfates

These elements can address the supply of sulfur, magnesium, and trace elements at the same time.

(1) Magnesium sulfate [$MgSO_4 \cdot 7H_2O$].

Magnesium sulfate appears as white crystals, contains 9.86% magnesium (Mg) and 13% sulfur (S), and is freely soluble in water: 35.5 g can be dissolved per 100 mL of water at 20°C. It is slightly hygroscopic and will cake after absorbing moisture. It forms a neutral aqueous solution and is a physiologically acidic fertilizer. Magnesium sulfate is the most commonly used magnesium source in soilless culture.

(2) Zinc sulfate [$ZnSO_4 \cdot 7H_2O$].

Commonly known as white vitriol, it appears as colorless rhombohedral crystals with a molecular weight of 287.55 and is freely soluble in water with 54.4 g dissolved per 100 mL of water at 20°C. It tends to lose crystalline water and turn into white powder under dry conditions. Containing 22.74% Zn, it is an important source of zinc nutrition for soilless culture.

(3) Copper sulfate [$CuSO_4 \cdot 5H_2O$].

It appears as blue crystals with a molecular weight of 249.69. Containing Cu 25.45% and S12.84%, it is freely soluble in water: 20.7 g can be dissolved per 100 mL of water at 20°C. Copper sulfate is a favorable source of copper nutrients for the soilless culture.

(4) Manganese sulfates [$MnSO_4 \cdot 4H_2O$ and $MnSO_4 \cdot H_2O$].

Manganese sulfates appear as pink crystals. The molecular weight of $MnSO_4 \cdot 4H_2O$ is 223.06 and it contains 24.63% of manganese; The molecular weight of $MnSO_4 \cdot H_2O$ is 169.01 and it contains 32.51% of manganese. They are both readily soluble in water.

7. Boron and Molybdenum Fertilizers

Boric acid, borax, ammonium molybdate, and potassium molybdate are mostly used.

(1) Boric acid [H_3BO_3].

It appears as white crystals with a molecular weight of 61.83. Containing 17.5% boron (B), it has low solubility in cold water (5 g boric acid dissolved per 100 g water at 20°C), and high solubility in hot water. It forms a slightly acidic aqueous solution and is a favorable source of boron for nutrient solutions for soilless culture.

(2) Borax [$Na_2B_4O_7 \cdot 10H_2O$].

It appears as white or colorless crystals with a molecular weight of 381.37 and a boron content of 11.34%. Borax tends to lose crystalline water and turn into white powder under dry conditions. It is freely soluble in water and is a favorable source of boron for nutrient solutions.

(3) Ammonium molybdate [$(NH_4)_6Mo_7O_{24} \cdot 4H_2O$].

It appears as white or light green crystals with a molecular weight of 1,235.86 and a molybdenum content of 54.38%; it is freely soluble in water and is a favorable source of boron for nutrient solutions.

(4) Potassium molybdate [K_2MoO_4].

It appears as white fine crystalline powder. With deliquescent and hygroscopic, it is freely soluble in water. Its molecular weight is 238.13, and it contains 40.31% of molybdenum. It dissolves easily in hot water and is a favorable source of boron for nutrient solutions.

(II) Physical Properties of Solid Substrates

The quality of the substrate is first determined by its physical properties. The fertility of substrate is not important in hydroponics. On the one hand, the substrate plays the role of fixing plants; on the other hand, it creates favorable hydroponic conditions for crop growth. Substrate cultivation requires the substrate to have excellent physical properties. The major indicators that reflect the physical properties of the substrate include bulk density, total porosity, void ratio, and particle diameter (particle size).

1. Bulk Density

Bulk density refers to the weight per unit volume of dry substrate, generally expressed in grams per liter (g/L) or grams per cubic centimeter (g/cm^3). Bulk density is different from specific weight, which is the mass per unit volume of solid substrate (with substrate porosity not calculated) and is based on the volume of the substrate itself. Bulk density is calculated in this way: Load the test substrate into a container of a known volume, pour it out to weigh it, and divide the weight of the substrate by the volume of the container.

The bulk density of the substrate is related to the particle diameter and total porosity of the substrate, and its size reflects the compaction, water-holding capacity, and air permeability of the substrate. High bulk density indicates that the substrate is too compact and not loose enough with excellent water-holding capacity but low air permeability. Low bulk density indicates that the substrate is too loose with high air permeability, which is conducive to the extension and growth of the root system, but with poor water-holding capacity, making it hard to fix the root system.

Different substrates vary greatly in bulk density (Table 5-5), and so does the same substrate considering the degree of compaction and particle size. For example, bagasse has a bulk density of 0.13 g/cm^3 when it is fresh and 0.28/cm^3 after nine months of stacking for decomposition. Generally, a low-bulk-density substrate has a bulk density of less than 0.25 g/cm^3; a medium-bulk-density substrate of 0.25–0.75 g/cm^3, and a high-bulk-density substrate of higher than 0.75 g/cm^3. Better effects can be achieved in crop cultivation if the substrate bulk density is in the range of 0.1–0.8 g/cm^3.

Table 5-5 Bulk Density and Specific Weight of Several Commonly Used Substrates

Substrate Type	Bulk Density (g/cm^3, approximate)	Specific Weight (g/cm^3)
Soil	1.10–1.70	2.45
Sand	1.30–1.50	2.62
Vermiculite	0.08–0.13	2.61
Perlite	0.03–0.16	2.37
Stone wool	0.04–0.11	—
Peat	0.05–0.20	1.55
Bagasse	0.12–0.28	—
Bark	0.10–0.30	2.00
Pine needles	0.10–0.25	1.90

It should be noted that the substrate can be measured from the perspective of dry bulk density and wet bulk density respectively. Assuming the dry bulk density of both perlite and vermiculite is 0.1 g/cm^3, after absorbing water, the former is twice its weight and the latter three times its weight. This means their wet bulk density is 0.2 g/cm^3 and 0.3 g/cm^3, respectively. In practice, wet bulk density may be more intuitive than dry bulk density. For example, the dry bulk density of artificial soil is 0.01 g/cm^3, which may cause an impression that the soil is too light to hold the plant roots in place. This, however, is not true as its wet bulk density can be up to 0.2–0.3 g/cm^3, close to that of perlite and vermiculite. Thus, no such misconception will be caused.

2. Total Porosity

Total porosity is the percentage of the substrate volume occupied by pore spaces, including both water-holding and air-filled pores. The substrate with high total porosity has more space to hold air and water, and vice versa. Total porosity can be calculated using the following formula:

$$\text{Total porosity} = (1 - \text{Bulk density/Specific weight}) \times 100\%$$

If a substrate has a bulk density of 0.1 g/cm^3 and a specific weight of 1.55 g/cm^3, the total porosity is $(1 - 0.1/1.55) \times 100\% = 93.55\%$.

High total porosity indicates light and loose substrate which is conducive to the growth of the root system but is not beneficial for fixation, making it easier for crops to fall over. For example, the total porosity of such substrates as bagasse, vermiculite, and rock wool is 90%–95% or higher. The substrate with a low total porosity is heavy, holding a small amount of water and air. For example, the total porosity of sand is about 30%. Therefore, in practice, to solve the problem that the total porosity of a single substrate is too high or too low, two or three substrates with different particle sizes are often mixed to make a composite substrate.

A substrate is classified as a medium-porosity substrate based on a percentage of macropores of 5%–30%, a low-porosity substrate based on a percentage of macropores of less than 5% and a high-porosity substrate based on a percentage of macropores of 30% (when the substrate shows low water-holding capacity and is easy to dry). In general, a proper substrate porosity is in the range of 54%–96%.

3. Void Ratio (air-water ratio of substrate)

Total porosity can only reflect the total void space occupied by both air and water in a substrate but cannot reflect the void space occupied by them respectively. How much air and easily available water can be supplied around the root system of growing plants is the most important physical property of horticultural substrates. The substrate with the best total porosity can provide 20% air and 20%–30% easily available water.

The air-water ratio refers to the relative ratio of air to water in the substrate for a certain period of time, which is usually expressed as the ratio of macropores to micropores in the substrate, and the macropore value is taken as 1. A macropore is a space occupied by water in the substrate, namely water-holding pores. The ratio of air-filled pores to water-holding pores is called the void ratio. It is expressed as the following formula:

$$\text{Void ratio} = \frac{\text{Air-filled pores (\%)}}{\text{Water-holding pores (\%)}}$$

Air-filled pores generally refer to those with a diameter of above 0.1 mm and serve to hold air. In these pores, the irrigating fluid will not be held but flows out under gravity. Water-holding pores generally refer to those with a diameter within 0.001–0.1 where water will be absorbed due to capillary action; hence they are also called capillary pores. Water present in these pores is called capillary water, and the main function of such pores is to store water rather than air.

The void ratio can reflect the distribution of air and water in the substrate and is an important indicator of substrate quality. It can comprehensively indicate the state of air and water in the substrate when used in combination with total porosity. A high void ratio means a high air-holding capacity and a low water-holding capacity. Generally speaking, a proper void ratio is in the range of 1 : 2–1 : 4, where the substrate has a high water-holding capacity and excellent air permeability, promising good crop growth and convenient management.

4. Particle Diameter (particle size)

The size of the substrate particles directly affects the bulk density, total porosity, and void ratio. Particle size is expressed as particle diameter (mm): The larger the particle of the same substrate, the higher the bulk density, the lower the total porosity, and the higher the void ratio. In contrast, the finer the particle, the lower the bulk density, the higher the total porosity and the lower the void ratio. Therefore, to meet the requirements of both water and oxygen absorption of the root system, the particles of the substrate should not be too large. Large particles indicate excellent air permeability but poor water-holding capacity, making it necessary to increase the watering frequency in planting management. Fine particles indicate a high water-holding capacity but poor air permeability, raising the possibility of excess moisture in the substrate, resulting in strong redox and affecting the root system growth. Therefore, the substrate particles should be moderate in size with rough but not angular surfaces and abundant and properly proportioned pores. The proper particle size varies among different types of substrates: It should be 0.5–2.0 mm for sand particles, within 1 cm for ceramsite, and whatever for substrates such as (blocky) stone wool as their particle size will not exert any of the above impacts. The physical properties of several commonly used substrates are listed in Table 5-6.

Table 5-6 Physical Properties of Several Commonly Used Substrates

Substrate	Bulk Density (g/cm^3)	Total Porosity (%)	Macropores [aeration volume(%)]	Micropores [capillary volume(%)]	Void Ratio [Macropores(%)/ Micropores(%)]
Vegetable garden soil	1.10	66.0	21.0	45.0	0.47
Sand	1.49	30.5	29.5	1.0	90.50
Cinder	0.70	54.7	21.7	33.0	0.64
Vermiculite	0.13	95.0	30.0	65.0	0.46
Perlite	0.16	93.2	53.0	40.0	1.33
Stone wool	0.11	96.0	2.0	94.0	0.02
Peat	0.21	84.4	7.1	77.3	0.09
Sawdust	0.19	78.3	34.5	43.8	0.79
Carbonized rice husk	0.15	82.5	57.5	25.0	2.30
Bagasse (stacking for 6 months)	0.12	90.8	44.5	46.3	0.96

When preparing a mixed substrate, the total volume of substrates with different particle sizes is smaller than that of raw materials. For example, after 1 m^3 of sand and 1 m^3 of bark are mixed, the total volume becomes 1.75 m^3 instead of 2 m^3 as the sand fills the pores of the bark. Moreover, this value will further decrease over time due to bark decomposition, which will weaken air permeability. Therefore, it is advisable to select organic substrates that are resistant to decomposition when preparing the mixed substrate so as to avoid the change of particle size from large to small over time. Compared with organic substrates, the particle size of inorganic substrates is not easily decreased through decomposition.

In addition, the substrate for cultivation should also be in good shape. Irregular particles have a larger surface area and can hold more water, while porous materials can also retain water inside the particles, thus holding more water.

III. Chemical Properties of Solid Substrates

The main chemical properties of the substrate that have a large impact on the growth of cultivated crops are the chemical composition of the substrate and the resulting chemical stability, acid-base properties, physical and chemical absorption capacities (cation exchange capacity), buffering capacity, and conductivity. Understanding the chemical properties and functions of substrates helps to well guide the selection of substrates and the preparation and management of nutrient solutions, thus improving cultivation management.

1. Chemical Composition and Stability of Substrates

The chemical composition of the substrate refers to the type and content of chemical substances contained, including organic and mineral nutrients that can be absorbed and utilized by plants, as well as toxic and harmful substances. The chemical stability of a substrate is its chemical resistance. Some substrates with low chemical resistance may produce some harmful substances after the chemical reaction, which not only harm the plant root system, but also disturb the original chemical equilibrium of the nutrient solution, and affect the effective absorption of various nutrients by the root system. Therefore, materials with strong stability should be selected as the substrate for soilless culture. Stable substrates mean little disruption of the chemical equilibrium of the nutrient solution and facilitate the daily management of the nutrient solution.

Different types of substrates have different chemical compositions, leading to different chemical stability. Generally speaking, substrates mainly composed of inorganic substances, such as river sand and gravel, have high chemical stability;

While those mainly composed of organic substances, such as wood chips and rice husks, have low chemical stability. However, peat shows stable chemical properties and is the safest to use.

The chemical stability of the substrates varies greatly with chemical composition. Substrates such as sand and gravel that are composed of inorganic minerals, such as quartz, feldspar, and mica, have the highest chemical stability, followed by substrates made of hornblende and pyroxene, among others. The most unstable substrates consist of carbonate minerals such as limestone and dolomite. In soilless culture, the former two types will not affect the chemical equilibrium of the nutrient solution by producing relevant harmful substances, but the latter will to a great extent by producing calcium and magnesium ions, which should always be noted.

Substrates composed of plant residues, such as peat, wood, rice husks, and bagasse, have complex chemical compositions and a great impact on the nutrient solution. Chemical composition can be roughly classified into three categories based on the impact on the chemical stability of the substrate: Categories I, II, and III. Category I substances are those susceptible to microbiological decomposition, such as sugar, starch, hemicellulose, cellulose, and organic acid in carbohydrates; category II substances are toxic, such as certain organic acids and phenols (tannin) and category III substances are those that are not easily decomposed by microorganisms, such as lignin and humus. Substrates containing Category I substances (such as fresh straw and bagasse) will cause strong biochemical reactions due to microbial activity in the early stage, which will seriously affect the equilibrium of the nutrient solution. The most noted consequence is a serious nitrogen deficiency. Substrates that contain many Category II substances can directly harm the root system. Therefore, the substrate with more substances of Categories I and II cannot be used directly without treatment. Substrates containing mainly Category III substances are the most stable and safest to use, such as peat, composted and decomposed wood chips, bark, and bagasse. Composting is the process of eliminating easily decomposed and toxic substances in the substrate, and transforming them into those that are difficult to decompose.The content of nutrient elements of common substrates is shown in Table 5-7.

Table 5-7 **Content of Nutrient Elements of Common Substrates**

Substrate	Total Nitrogen (%)	Total Phosphorus (%)	Available Phosphorus (mg/L)	Available Potassium (mg/L)	Available Calcium (mg/L)	Exchangeable Magnesium (mg/L)	Available Copper (mg/L)	Available Zinc (mg/L)	Available Iron (mg/L)	Available Boron (mg/L)
Vegetable garden soil	0.106	0.077	50.0	120.5	324.70	330.0	5.78	11.23	28.22	0.425
Cinder	0.183	0.033	23.0	203.9	9247.5	200.0	4.00	66.42	14.44	20.3
Vermiculite	0.011	0.063	3.0	501.6	2560.5	474.0	1.95	4.0	9.65	1.063
Perlite	0.005	0.082	2.5	162.2	694.5	65.0	3.50	18.19	5.68	—
Stone wool	0.084	0.228	—	1.338*	—	—	—	—	—	—
Cotton seed shells	2.20	0.210	—	0.17*	—	—	—	—	—	—
Carbonized rice husks	0.54	0.049	66.0	6625.5	884.5	175.0	1.36	31.30	4.58	1.290

Note: * refers to the percentage of total potassium (%).

2. Acid-base Property (pH) of Substrate

The pH level indicates the acid-base property of the substrate: pH=7 is neutral, pH<7 is acidic, and pH>7 is alkaline. A change of one pH unit corresponds to a 10-fold change in the power of hydrogen. For example, there is a 10-fold increase in acidity from pH 5 to pH 6 and a 100-fold increase in acidity from pH 5 to pH 7.

The substrate itself has a certain acidity or alkalinity. Substrates with too high acidity or alkalinity will affect the acidity or alkalinity of the nutrient solution, seriously disrupting the chemical equilibrium of the nutrient solution and hindering nutrient absorption of plants in severe cases. Therefore, a general understanding of the pH of a substrate is required before it is selected so as to take corresponding measures to regulate the pH level. A simple method to test the pH level of a substrate is to take 1 portion of the substrate, add distilled water 5 times its volume, mix them well, and then test the pH value with test paper or an acidimeter.

Although most ornamental plants are adaptable in the pH range of 5.5–6.5, a proper pH level of a substrate should be in the range of 6.5 (slightly acidic)–7.0 (neutral) to facilitate easy artificial regulation and avoid any influence on the effects of certain components in the nutrient solution and subsequent physiological disorders of the plant after the nutrient solution is supplied.

Calcareous gravel contains abundant calcium carbonate. It dissolves in the nutrient solution, which raises the pH level and causes iron precipitation, resulting in iron deficiency in plants. Therefore, it is not suitable to be used as a substrate. For furfural of high acidity, its pH level must be adjusted to make it slightly acidic with alkaline substances; otherwise, it should not be used as a substrate.

In general, as most nutrient solutions are acidic, after the substrate is supplied with a nutrient solution several times, its pH level will slightly drop or become close to that of the nutrient solution. If the acidity of the substrate is adjusted with alkaline substances, trace element deficiency may be caused.

3. Cation Exchange Capacity (CEC)

The CEC of the substrate is expressed as the milligram equivalents (mEq/100 g of the substrate) of absorbed cations exchanged per 100 g of substrate. The cation exchange capacity can not only reflect the capacity of the substrate to absorb and retain fertilizer nutrients and to protect the fertilizer ions from being leached by water and slowly release them for plant absorption, but also can buffer the acid-base reaction of the nutrient solution. Some substrates have almost no cation exchange capacity (such as most inorganic substrates) and some have strong cation exchange capacity, which has a significant impact on the composition of the nutrient solution in the substrate. Substrates with high base exchange capacity show strong nutrient retention, but a too-high cation exchange capacity will result in the accumulation of soluble salts, which is likely to cause harm to plants due to difficulty in nutrient leaching. In contrast, a too-low cation exchange capacity will lead to the retention of only a small amount of nutrients, thus requiring frequent fertilizer application. Substrates with high base exchange capacity can mitigate rapid pH changes in nutrient solutions but require the use of a large amount of buffer to adjust the pH level. In general, organic substrates with high base exchange capacity have high buffering capacity, which helps to resist nutrient leaching and excessive pH changs.

The cation exchange capacity of the substrate has its downside, that is, affecting the equilibrium of the nutrient solution and making it difficult to control its composition as needed. There is also an upside, that is, reducing the loss of nutrients and buffering the acid-base reaction of the nutrient solution. To select a substrate, it is necessary to know its cation exchange capacity to weigh the pros and cons. The cation exchange capacity of several commonly used substrates is listed in Table 5-8 to illustrate the differences.

Table 5-8 Cation Exchange Capacity of Several Commonly Used Substrates

Substrate Type	Cation Exchange Capacity (mEq/100g)
High-lying peat	140–160
Intermediate peat	70–80
Vermiculite	100–150
Bark	70–80
Inert substrates such as sand, gravel, and stone wool	0.1–1

In pot planting, the cation exchange capacity is generally expressed as milligram equivalents of cations adsorbed per 100 g. Normally, a base exchange capacity of 10–100 mEq/100 g is considered suitable, less than 10 mEq/100 g is considered low, and higher than 100 mEq/100 g is considered high.

4. Conductivity of Substrate

The conductivity of substrate refers to the conductivity of the substrate itself before the nutrient solution is added. It can be determined with a conductivity meter, and reflects the amount of soluble salt originally contained in the substrate, which will directly affect the equilibrium of the nutrient solution. For example, sand affected by seawater often contains a high salt content and should be properly treated. The soluble salt content in the substrate should not exceed 1,000 mg/kg and preferably not exceed 500 mg/kg. The conductivity of the substrate should be determined before being used to facilitate leaching with freshwater or other appropriate treatment.

Based on the correlation between the conductivity of the substrate and nitrate nitrogen, the conductivity value can be used to infer the nitrogen content of the substrate and determine whether nitrogen fertilizer is required. Generally, in flower cultivation, fertilizer must be applied when the conductivity is lower than 0.37–0.5 ms/cm (equivalent to that of tap water); fertilizer is generally not applied when the conductivity reaches 1.3–2.75 ms/cm and salt leaching should be carried out where possible. The conductivity for cultivating vegetable crops should be greater than 1 ms/cm.

5. Buffering Capacity of Substrates

The buffering capacity of a substrate refers to its ability to buffer pH changes after fertilizer is applied. The buffering capacity is mainly determined by base exchange capacity and the content of weak acids and salts present in the substrate. Generally, high base exchange capacity means high buffering capacity. Substrates containing rich calcium carbonate and magnesium salts show strong buffering capacity for acid but no buffering capacity for alkali; substrates containing rich humus show buffering capacity for both acid and alkali. Substrates ranking by buffering capacity are organic substrate > inorganic substrate > inert substrate > nutrient solution. Some of the commonly used substrates , such as vermiculite, show strong buffering capacity. Nevertheless, most of them have weak buffering capacity. Therefore, it is required to understand the buffering capacity of substrates so as to make the best use of their advantages and avoid their disadvantages.

6. Carbon-Nitrogen Ratio

It is the relative ratio of carbon to nitrogen in the substrate. Substrates with a high carbon-nitrogen ratio (high carbon content to low nitrogen content) may cause nitrogen deficiency of plants as microorganisms compete for nitrogen during vital activities. Substrates with a high carbon-nitrogen ratio disallow plants to grow normally or develop properly, even if good cultivation technologies are adopted. Therefore, the dosage of organic substrates such as wood chips and bagasse should not exceed 20% to prepare a mixed substrate, or 8 kg of nitrogen fertilizer should be added per m^3. The substrate should not be used before being composted for 2–3 months. In addition, the organic substrate with large particles has a slow decomposition rate as its surface area is smaller than its volume, and its effective carbon-nitrogen ratio is lower than that of the organic substrate with fine particles. Therefore, substrates with coarse particles, especially those with a low carbon-nitrogen ratio, should be used where possible.

According to general provisions, a carbon-nitrogen ratio of 200 : 1–500 : 1 is considered medium; lower than 200 : 1 is considered low; higher than 500 : 1 is considered high. Generally, the carbon-nitrogen ratio should be medium to low rather than high. C : N=30 : 1 or so is more suitable for crop growth.

IV. Case Sharing

Growing Vegetables in the Desert with New Soilless Culture Technology

In Kuwait, the temperature soars to 60°C or even higher in summer, and almost no grass grows in such a hot and dry environment, making Kuwait almost completely dependent on imported food. Fortunately, new technology has made eating locally cultivated vegetables no longer a luxury for Kuwaitis.

Dr. Glenn Carrigan, an agricultural expert from Australia, and Dr. Abdulla from Kuwait successfully helped vegetables to survive the local heat with novelty soilless culture technologies. Vegetables are placed in special soilless containers in greenhouses. The special shade covering these containers can reflect about 50% of the sunlight, lowering the temperature by about 10°C and greatly reducing the solar heat on vegetables.

"We also raised the humidity in the air," said Dr. Glenn Carrigan, " To lower the temperature since the normal humidity in the desert in summer is 7%–11%, which is quite dry. Therefore, by increasing water vapor in the air, the temperature drops from 40°C or 50°C to 20°C or 30°C, which is beneficial to the growth of vegetables." Water for humidification is collected in the water tank for recycling, which is essential in Kuwait where water is an expensive resource. At present, the water used on the farm is taken somewhere 40 m deep underground and is usually slightly saline. "Water is valuable in Kuwait," said Dr. Abdulla, "To improve quality, water is thoroughly purified to obtain freshwater which hopefully can be used for fish farming in the near future. Vegetables can be planted outdoors in winter and transferred into the greenhouse when the temperature rises in March. These vegetables grow in an almost sterile environment, so no pesticide is required."

This is the first farm in Kuwait to adopt soilless culture technology and vegetables such as lettuce and celery are planted in a greenhouse of 2,000 m^2. Freshly picked vegetables are supplied to customers within 3 hours every day. Dr. Abdulla hopes to expand the scale as soon as possible, increasing the yield 3–4 times.

Module 6 Construction and Environmental Regulation of Protection Facilities

[Learning Objectives]

I. Knowledge

(1) Understand the origin and evolution of lettuce;

(2) Master the botanical characteristics of lettuce and its required environmental conditions during different development periods.

II. Skills

(1) Learn to retrieve relevant information;

(2) Learn about the required environmental conditions in different growth stages of lettuce.

[Preparation]

I. Required Resources

(1) A paper library and a periodicals reading room;

(2) A digital reading room and an electronic resource library;

(3) A multimedia classroom.

II. Background Knowledge

(1) Master the basic methods of literature review;

(2) Understand the relevant knowledge of lettuce botany;

(3) Visit demonstration bases of lettuce cultivation and production.

[Learning Tasks]

Task 1 Type and Classification of Protection Facilities

An environmental protection facility refers to a cultivation facility with a transparent covering that regulates ecological factors such as temperature, light, water, and air and allows agricultural work inside. Protection facilities are generally classified into glass greenhouses and plastic greenhouses by covering material. Plastic greenhouses can be further classified into hard-plastic (such as PC, FRA, FRP, and composite plates) and soft-plastic (such as PVC, PE, and EVA film) greenhouses according to the type of plastic; into single-span and multi-span greenhouses by span; into double-roof, single-roof, uneven-double-roof, and arched-roof greenhouses by roof.

The currently commonly used environmental protection facilities in China mainly include solar greenhouses, plastic greenhouses, modern greenhouses, canopies, and sunshade nets. Plastic greenhouses have been widely used throughout China, especially in southern China, where plastic greenhouses are developing towards tall multi-span greenhouses on a large scale. Solar greenhouses are primarily promoted and applied in the northern region of 33° N–46° N and the alpine region of -15– -20°C, where production of warm-season fruits and vegetables in winter is basically realized without heating, writing a brilliant chapter in the history of facility horticulture in China. Since the 1990s, large modern greenhouses have been introduced from other countries. Through learning, digestion, and absorption, China has developed its own modern greenhouses, which have promoted the scientific and technological progress of greenhouse horticulture and the modernization of agriculture in China, offering great opportunities for the development of soilless culture technology. Summer protection facilities, such as canopies and sunshade nets, play an important role in overcoming the threat of catastrophic climate and adverse environments such as frequent high temperatures, rainstorms, typhoons, diseases, and pests in hot summer in southern areas, and provide a simple and effective new way to alleviate the shortage of horticultural products in off seasons (summer and autumn) in southern areas. Due to the different natural, climatic, economic, technical, labor, and other conditions in different places, facilities vary in type, structural material, and scale, and should be selected and sited based on local conditions.

I. Plastic Greenhouse

Generally, large protective cultivation facilities with plastic film covering and bamboo, wood, concrete, or steel-pole frameworks are called plastic film greenhouses, or plastic greenhouses for short (Fig. 6-1). According to roof form, plastic greenhouses can be divided into arched and ridge-shaped, with the former mostly seen. According to the framework material, plastic greenhouses can be classified into bamboo-structured, wood-structured, steel-frame (pipe)-structured, steel-bamboo-structured, reinforced-concrete-structured, and inflation-type greenhouses. According to the connection method, they can be divided into single-span, double-span, and multi-span greenhouses. A plastic greenhouse has a simple structure that is mainly composed of arch frames, longitudinal beams, upright columns, gable wall columns, framework connecting clamps, and doors and it varies according to construction materials. The arch frame is the main component of plastic greenhouses that can withstand wind, snow, and other loads.

Fig. 6-1 Plastic Greenhouse

It mainly comes into two types: single-rod type and truss type. The longitudinal beam is a component that ensures the longitudinal stability of arch frames by connecting them into one. It also comes into two types: single-rod type and grid type. When the cross-section of the arch frame is small and cannot withstand wind and snow loads, or the span of the arch frame is large, and the strength of the supporting structure is insufficient, upright columns should be set up in the greenhouse to directly support the arch frames and longitudinal beams to improve the overall bearing capacity of the plastic greenhouse. Gable wall columns, i.e. head wall columns, are commonly upright and arched or diagonal in areas with strong wind. Special prefabricated clamps are used for connections between the assembled plastic greenhouses made of galvanized steel pipes and the RC-structured plastic greenhouses, such as the connections between the arch frame and the gable wall column, between arch frames, and between the longitudinal beam and head wall arch, The plastic greenhouses of a bamboo or wood structure are connected by threads, ropes, or iron wires. The door of the plastic greenhouses is designed not only for management and transportation but also for ventilation. It is generally located in the middle of the head wall of a single-span greenhouse. For thermal insulation, the door can be set at the southern head wall. After the temperature rises, another door can be set at the northern end to strengthen ventilation. Such doors can be hinged doors, hanging-track sliding doors and the like. Plastic greenhouses for soilless culture should have a solid structure with strong resistance to wind and snow, adequate space for plant growing and a large area, and feature easy ventilation and environmental regulation. Preferably, a greenhouse should have a steel structure without or with few pillars and should span 8–12 m with a length of 40–60 m and a ridge height of 2.4–3.0 m. At present, assembled greenhouses made of galvanized thin-wall steel pipes are generally used in production as they are of a uniform specification, need no upright columns for construction, and are easy to assemble and disassemble, offering large space for agricultural work and making it convenient to fix the film covering.

In terms of performance, plastic greenhouses are well-sealed and receive sufficient light in a uniform manner: short-wave radiation is easy to penetrate, while long-wave radiation is difficult to penetrate. Greenhouses of a single-span steel structure or a hard-plastic structure are well-lighted, while those of a single-span bamboo or wood structure or multi-span structure are limitedly-lighted. In addition, the larger the span of the greenhouse, the higher the greenhouse frame and the weaker the light in the greenhouse. On sunny days, the greenhouse experiences a rapid temperature rise during the daytime and preserves certain heat at night. However, due to the lack of night heating measures and great daily and seasonal temperature variations, most greenhouses with no essential environmental regulation equipment have to rely on ventilation windows to adjust the temperature and humidity, resulting in great variations of indoor environmental elements. On sunny days, the temperature in the greenhouse rises rapidly 1–2 hours after the sun comes out and climbs linearly at a rate of 5–8°C per hour during 8:00–10:00. During 11:00–13:00, the temperature rise slows down and reaches a summit, which is 20°C higher than the outdoor temperature. Then the indoor temperature begins to drop and drops rapidly from 15:00–17:00 at a rate of 5–6°C per hour. With the narrowing of the indoor and outdoor temperature difference, the nighttime drop rate will rapidly decline to about 1°C per hour. Early in the morning, the indoor temperature is only 3–5°C higher than the outdoor temperature. Sometimes the indoor temperature is lower than the outdoor temperature, which is called "temperature inversion", and mostly occurs on a sunny, breezy morning in early spring or late autumn. As the greenhouse is well sealed by the plastic film, the humidity indoors is significantly higher than that outdoors, with an average daily increase of 35%–40%. The northern area with marked seasonal changes has witnessed slow development of plastic greenhouses because of a long low-temperature period in winter and limited utilization (only about one month earlier or later than in the open air). In contrast, plastic greenhouses are widely applied in the southern warm areas or areas with a mild climate.

II. Solar Greenhouse

A solar greenhouse (Fig. 6-2) belongs to the single-roof greenhouse, commonly known as a "winter greenhouse". It is a horticultural protection facility

Fig. 6-2 Solar Greenhouse

with three walls, an over-2 m-high ridge, and a span of 6–8 m that mainly relies on solar radiation energy as the heat source (including at nighttime). Most solar greenhouses are covered with plastic film for lighting. Facing south and extending from east to west, they feature a large lighting surface and angle of incidence with solar radiation as the main heat source. In addition to the maximum lighting, the solar greenhouses are also provided with a wide range of lighting, thermal insulation, and cold-proof equipment and facilities, such as thickened walls and back slopes, as well as cold-proof ditches, paper quilts, and straw mats to achieve the effects of heating and heat preservation, thus making full use of photothermal resources and weakening the impact of unfavorable meteorological factors. Solar greenhouses are generally not heated or are only occasionally heated as a supplementary measure. The first-generation solar greenhouses are mostly constructed with bamboo/wood and earth walls, which are of low cost. The (front, middle and rear) pillars inside are generally arranged every 3 m, and the space between them is called a partition. The second-generation solar greenhouses are mostly made of steel and brick walls, without pillars inside and feature enlarged space, better operability, firm structure, improved resistance to wind and snow, longer service life, higher cost, and lower depreciation cost compared with the first-generation ones. Some may also have a steel-bamboo mixed structure, a GRC framework structure, or other structures. Energy-saving solar greenhouses covered by plastic film are the main form of vegetable protection facilities in northern China. They exhibit evident advantages in lighting, heat preservation, low-energy consumption, and practicality, with lower investments and higher benefits.

III. Modern Greenhouse

In facility horticulture, modern greenhouses (Fig. 6-3) are large, high-end greenhouses suitable for crop growth and development through computerized automatic control of various environmental elements in the facility. They are basically not affected by catastrophic weather and unfavorable environmental conditions and can produce horticultural crops all year round. Modern greenhouses have a service life of 20–50 years and are mostly of a truss structure made of hot-dip galvanized steel or aluminum profiles with a concrete foundation. Generally, they have a large lighting area that allows for a large amount of light to enter in winter in the east-west direction and a uniform temperature and light distribution in the south-north direction. If plastic greenhouses and solar greenhouses are protection facilities that "rely on natural conditions", modern greenhouses are protection facilities that "create natural conditions". In addition to the main structure, there are various environmental regulation devices inside, including heating systems, heat preservation systems, cooling systems, humidification or dehumidification systems, CO_2 application systems, forced ventilation systems and natural ventilation systems, light supplementation or shading systems, fertilizer and water supply systems, drainage and rainwater collection systems, protection systems (fly nets, snow removal equipment), weather stations, power systems and control systems. Cost varies greatly due to the adoption of different systems and equipment models. Expanding the greenhouse scale is the main way to reduce the cost of equipment per unit area, and connecting single-span greenhouses is an effective way to achieve it. Using large greenhouses will be the trend.

Fig. 6-3 Modern Greenhouse

At present, China's introduction of modern greenhouses mainly includes multi-ridge, multi-span, and small-roof Venlo glass greenhouses that were created in the Netherlands and then become popular worldwide; popular Richel plastic greenhouses created by Richel Group, France; full-open, multi-span, curved-roof plastic greenhouses that allow the film covering the top roof to be rolled up from bottom to top for ventilation; full-open glass greenhouses (future greenhouses) created by Serre Italia. Typical modern greenhouses with an automatic control system designed and manufactured in China include double-layer inflatable multi-span plastic greenhouses, double-sloped glass greenhouses, North China large multi-span plastic greenhouses, South China large multi-span plastic greenhouses, gold-topped multi-span greenhouses, LGP-732 multi-span greenhouses, XA and GK greenhouses, and fiber reinforced plastic (FRP) multi-span greenhouses.

IV. Plant Factory

A plant factory (Fig. 6-4) is a factory-like, fully enclosed facility in which plants are cultivated with higher efficiency, less labor, and greater stability using artificial light sources to realize automated control of the environment inside. Plant factories are divided into three categories according to different light sources: artificial light type, natural light type, and combined type. As part of the "controlled-environment agriculture", plant factories are the highest level of horticultural protection facilities, and their management is highly mechanized and automated. Soilless culture and three-dimensional planting of crops are carried out in large facilities. With changes in environmental elements transmitted to the computer through sensors, the operation of various environmental regulating equipment and the production process can be accurately controlled through calculation. The required temperature, moisture, light, water, fertilizer, air, and the like are optimally configured according to the requirements of plant growth. In addition to computer monitoring, fully enclosed production management is carried out by robots and manipulators to realize streamlined production from sowing to harvest, completely removing the natural constraints. This is to make rational use of space and facilities to achieve economical and efficient production as well as all-weather, seasonless, and pollution-free production. Plant factories are characterized by high integration, efficient production, high merchantability, and high investment. Based on biotechnology, engineering technology, and system management, they represent an industrialized agricultural system that produces plant products as planned on an annual basis. They also represent one of the most dynamic and promising fields that apply high-tech achievements in the process of agricultural industrialization, guiding the development direction of future agriculture.

Fig. 6-4 Plant Factory

V. Canopy and Sunshade Net

Abundant summer rainfall, strong light exposure, high-temperature stress, and frequent diseases and pests inhibit the growth of crops such as vegetables, resulting in unstable production. The simplest canopy cultivation is only to cover the top (canopy) and remove the side film (apron) of the greenhouse framework so that it can both prevent rain and realize ventilation in summer. If the canopy is then covered with a silver-gray or black sunshade net to resist strong light exposure, the substrate temperature in the greenhouse at noon in summer can drop by 8–12°C, effectively reducing the damage of high temperature.

Task 2 Construction of Plastic Greenhouses and Solar Greenhouses

I. Structure of Plastic Greenhouses

The framework of a plastic greenhouse consists of an upright column, an arch bar (arch frame), a pull rod (longitudinal beam/overhanging beam), and a pressure bar (tension rope), which is commonly known as "three bars and a column". It is the most basic framework of a plastic film greenhouse. Other forms have evolved on this basis.

1. Upright Column

The upright columns mainly function as support for the arch bar and greenhouse surface and are arranged in a straight line vertically and horizontally. Upright columns are mainly made of bamboo poles, cement prefabricated columns, and steel frames. Too many upright columns in the greenhouse will increase the shaded area and affect the work in the greenhouse. Therefore, "suspension of beam and column" can be used. That is, the longitudinal upright columns will be reduced and replaced by small davits fixed on the pull rods. The small davits are about 30 cm high from the ground and are arranged at an interval of 0.8–1 m on the pull rods, which is consistent with the spacing of the arch bars. Generally, the upright columns can be reduced by 2/3.

2. Arch Bar

Forming the framework of the plastic greenhouse, arch bars determine the shape and spatial composition of the greenhouse and support the film covering of the greenhouse. They are mainly made of bamboo poles, steel beams, steel pipes, and hard plastic pipes.

3. Pull Rod

Pull rods are mainly used to connect the arch bars and upright columns longitudinally, fix the pressure bars, and make the greenhouse framework a whole. Pull rods for a bamboo or wood structure are usually fixed at the upper part of the upright column (20–30 cm from the top), while those for a steel structure are generally fixed directly to the arch bars with their length consistent with that of the greenhouse body. Pull rods are mainly made of bamboo poles, steel beams and steel pipes.

4. Pressure Bar

The pressure bars are located in the middle of the two arch frames above the film and play the role of flattening, compacting, and tightening the film. The pressure bar can be made of smooth and straight bamboo poles or replaced with galvanized wires or nylon ropes. At present, a special plastic tension rope is available to replace the pressure bar.

II. Construction of Plastic Greenhouse

A proper type of plastic greenhouse should be selected according to the actual local conditions and by following the principle of using local materials, high cost-efficiency and practicality, convenient management, and maximum profit. At the same time, the construction height, width, and length of the greenhouse should be considered, together with its robustness and thermal insulation.

The length of the plastic greenhouse should be 40–60 m and should not exceed 100 m at the maximum. Being too long may cause inconvenience for production and management. The span is mostly 8–12 m and a span that is too wide will cause poor ventilation in the greenhouse. Under the same conditions (including with the same area), the larger the span of the greenhouse, the greater the load on the arch bar, the lower the wind resistance, and the more difficult in installing the plastic film. It will be hard to stretch the film tightly, which will shake the film more frequently and make it more susceptible to wind damage. Conversely, the smaller the span, the stronger the wind resistance. The aspect ratio of the greenhouse has a great impact on stability. For greenhouses with the same area, the larger the aspect ratio, the longer the circumference; the more the fixed parts on the ground, the stronger its stability; the smaller the span, the smaller the effective utilization area. It is generally considered that an aspect ratio of 5 or higher is better.

The greenhouse should not be too high, as the higher the greenhouse, the more the speed increases when the airflow passes, and the greater the change in different parts, the more sever the oscillation of the film. Therefore, the greenhouse should be as low as possible on the premise of meeting the requirements for space. Practice has proved that the height-to-span ratio (height/span) of northern greenhouses is better at 0.25–0.30 while that of southern greenhouses should be larger at 0.3–0.4 considering natural ventilation and other factors. In addition, the shape of greenhouses should be streamlined as the use of a linear surface enables stable airflow when the wind blows and reduces the oscillation of the film. Evenly tensioned film covering is conducive to improving its wind resistance.

III. Structure of Solar Greenhouses

A solar greenhouse consists of a rear wall, rear roof, front roof, and bilateral gable walls. The length, width, size, thickness, and materials used for each part determine the lighting and heat preservation of the solar greenhouse.

1. Angle

It refers to the front roof angle, rear roof angle, and azimuth angle. The front roof angle is the angle between the bottom of the front roof of the greenhouse and the ground level, and it determines the lighting of the greenhouse. The angle should allow sunlight to enter the greenhouse to a maximum extent during winter and spring and ensure a temperature of above 20.5°C in the greenhouse in areas at 32°N–43°N. In addition, the rationality of the overall structure, shape, service area, and working space of the greenhouse should also be considered, so the angle between the tangent of the ground circle and the front roof of the solar greenhouse should be 60°–68°. The rear roof angle refers to the angle between the inner side of the rear roof and the ground level, which should be 5°–8°, greater than the local solar height angle at noon on "winter solstice". Azimuth angle refers to the orientation of a solar greenhouse, which requires the solar greenhouse to face south and extend from east to west, and appropriately deviate towards east or west by 5°–10° based on the actual local situation to make full use of the light in the morning or afternoon.

2. Height

It refers to the ridge height and rear wall height. Ridge height is the vertical distance from the greenhouse ridge to the ground. The ridge height and the span are in a certain relationship. In the case of a constant span, reducing the ridge height will decrease the front roof angle and indoor space of the greenhouse, resulting in lessening incident light and hindering crop growth and indoor agricultural work. Increased ridge height can enlarge the front roof angle and expand the greenhouse space, which is conducive to lighting. However, a ridge being too high will increase the construction cost and affect heat preservation due to an excessively large heat dissipation area.

3. Span

It refers to the distance from the inner side of the north wall of the greenhouse to the south base angle of the transparent roof. The span of the greenhouse has a great impact on its lighting, heat preservation, crop growth and development, and agricultural work. The span is restricted by height. Under a certain greenhouse height, expanding the span will result in a reduction of the front roof angle and indoor space, which is not conducive to lighting and agricultural work. In addition, the increased heat dissipation area is not beneficial for heat preservation at night. In contrast, a small span leads to a low land utilization rate.

4. Length

It refers to the distance between the east and west gable walls and is preferably 50–60 m. A small length will increase the cost per unit area, and the ratio of the shaded area of the east and west gable walls before sunrise and sunset to the greenhouse area, and reduce the greenhouse volume, thus affecting the absorption and release of heat. Therefore, the length of the greenhouse should not be less than 30 m. However, a large length will cause difficulties in regulation and vulnerability to wind damage, so the maximum greenhouse length should not exceed 100 m.

5. Thickness

It refers to the thickness of the walls and rear roof. The greenhouse wall and rear roof are designed for heat preservation and storage, so the heat conductivity, heat storage coefficient, and sufficient thickness of building materials should be considered. In general, heterogeneous materials with a high inner-layer heat storage coefficient and a low outer-layer heat conductivity are preferred for greenhouse walls. For earth, stone, or brick walls, the total thickness should be the thickness of the local frozen soil layer plus 50 cm. The rear roof should be made of materials such as stover, straw, or wood boards plus polystyrene boards, slag, and cement. The total thickness should be 40–70 cm.

6. Front-to-Rear Slope Ratio

It refers to the ratio of the vertical projection width of the front and rear slopes. Considering heat preservation, lighting, convenient operation, and expansion of cultivation area, the front-to-rear slope ratio should be about 4.5 : 1. For example, for a greenhouse with a span of 6–7 m, the projection width should be 5–5.5 m for the front roof and 1.2–1.5 m for the rear roof.

7. Height-to-Span Ratio

It refers to the ratio of ridge height to the span of the solar greenhouse. The height-to-span ratio determines the roof angle, which is preferably 1 : 2.2. For example, for a greenhouse with a span of 6 m, the height should be above 2.7 m and for a greenhouse with a span of 7 m, the height should be above 3 m.

8. Thermal Insulation Ratio

It refers to the ratio of heat storage area to heat release area in the solar greenhouse. In the solar greenhouse, the heat storage surface is the ground inside the greenhouse, and the heat dissipation surface is the front roof, i.e. thermal insulation ratio (R) of the solar greenhouse = land area in the solar greenhouse (S)/ front roof area of the solar greenhouse (W).

The thermal insulation ratio (R) indicates the thermal insulation performance of the solar greenhouse. The higher the thermal insulation ratio, the better the insulation performance. If the thermal insulation ratio is equal to 1, i.e. the land area is equal to the front roof area, the thermal insulation effect is the best.

9. Shade Ratio

It refers to the ratio of the distance between the middle pillar of the front greenhouse and the front edge of the rear greenhouse to the ridge height or ridge height plus the diameter of the straw mat (thermal insulation quilt) of the greenhouse when building multi-span greenhouses or building greenhouses on the north side of a tall building where the front greenhouse does not affect the lighting of the rear one greenhouse. The distance from the middle pillar of the front greenhouse to the front edge of the rear greenhouse (greenhouse spacing) is equal to the ridge height or ridge height plus the diameter of the straw mat (thermal insulation quilt) of the greenhouse multiplied by cotangent of the local solar height angle at noon of the Winter Solstice. The practice has proved that a suitable shade ratio should be greater than 2.5 ∶ 1.

IV. Construction of Solar Greenhouse

1. Site Positioning, Ground Leveling and Setting Out

Site positioning is the process of determining the location and directions of paths and boundaries within the site based on the design drawing. The orientation of the solar greenhouse should be east-westward, with the lighting surface facing south. It is advisable to measure the magnetic meridian by compass, then adjust and measure the true meridian according to the local magnetic declination, and measure the east-west direction of the vertical path. Alternatively, the upright pole method can be adopted to determine the north-south direction: Erect a wooden pole vertically to the ground at the proposed site, and measure the shadow length and location of the wooden pole every 10 minutes from 10:00 to 14:00. Among the shadow length values, the shortest one represents the north-south direction. Then determine the east-west direction by the Pythagorean theorem: Measure 3 m along the north-south direction and measure 4 m eastward with a meter ruler so that the distance between this point and the wooden pole is 5 m, forming a right triangle. The 4 m extension line is due east-west direction. After the paths in the site are positioned, the construction land of the greenhouse should be leveled and cleared of debris, and then key points should be set out according to the structural parameters of the greenhouse.

2. Greenhouse Foundation

The foundation is one of the important structures of the greenhouse and bears all the loads of the upper part. Therefore, the bearing capacity of the foundation should be guaranteed to avoid excessive deformation. Foundations should be built for masonry or earth walls, and their depth should be greater than the maximum depth of the local frozen soil layer, considering the impact of soil freezing and thawing on the foundations. The wall base should be compacted for mechanically compacted walls. Bricks, rubbles, or concrete are generally used as foundation materials.

3. Wall

(1) Earth wall: The rammed earth walls and straw-mud walls have been basically abandoned because they are too labor-intensive. Mechanically compacted walls are commonly used, i.e. excavators and chain-link bulldozers which are used together for construction to transfer indoor mellow soils to the south side. Then, the soil is collected by a bulldozer and stacked at the wall, and repeatedly compacted by a bulldozer every time the stacked height reaches 0.4–0.5 m until the required height is reached. The rear wall is then cut with an excavator to create a certain slope. Alternatively, a built-in protective brick wall is built to prevent the wall from sliding and collapsing.

(2) Brick wall: The thickness of brick walls is determined according to the insulation requirements. However, considering the high construction cost, a solid brick wall cannot be built too thick. As a result, heterogeneous composite walls are

generally built. There are no tension bricks between the two layers of brick walls, but insulation materials such as perlite, vermiculite, and slag. The brick walls are well sealed to prevent frost heave caused by water ingress. When constructing the rear wall, the outer side should be raised by 30–40 cm to avoid an overlarge slope of the rear layer surface. The sidewall should be in the same shape as the front roof.

4. Rear Roof

The rear roof framework of the solar greenhouse with a bamboo-wood structure is mainly composed of middle pillars, girders, and purlins. The solar greenhouse with a steel structure may be built with no middle pillars. As the main load-bearing member of the entire rear roof, the middle pillars must be solid, durable and padded with bedrock or poured with concrete to prevent sinking during use. The middle pillars shall be tilted 4°–5° northward to increase their stability and enable them to withstand the horizontal thrust of the rear roof well. The girder is placed on the middle pillars, with its head extending above the middle pillars by 20–30 cm and its tail on the rear wall. Fixing measures shall be taken to prevent them from sliding. A total of 3–4 ridge purlins are set on the girder head. Wire ropes may be used in place of them. Wooden boards, prefabricated slabs, or waste plastic film can be laid on the purlins or wire ropes, and then stover, straw, slag, and other thermal insulation materials can be laid. Where possible, a polystyrene board with a thickness of 5–6 cm can be added, coated with cement on the surface, and covered with asphalt felt for waterproofing.

5. Front Roof

The front roof is composed of upright columns, cross beams, an arch frame, plastic film, and tension ropes. The front roof framework used to bear the covering, wind and snow loads must have sufficient strength. The section of the framework should be reduced where possible without affecting its strength so as to reduce the shade. According to the types of greenhouses, 1–2 rows of upright columns made of wood or cement concrete can be provided to support the front roof loads. Suspension of the beam and column may also be used to reduce the upright columns and enhance operability. The upper and lower ends of the steel frame greenhouse may be directly fixed on the embedded parts of the rear wall and the southern grade beam without setting upright columns.

The greenhouse should be covered with mulch film on windless sunny mornings before frost sets in. For solar greenhouses whose front roofs are different in width, the number of film strips required is different; generally, 2–3 strips are required. The lower film strip is 1–1.5 m wide, with its lower end buried in the soil or fixed with a circlip; the middle film strip varies in width depending on the span of the greenhouse; the upper film strip is also 1–1.5 m, with its upper end fixed on the rear roof. The upper end of each film strip should be bonded with a stronger 20 cm-wide strip, with a hemp rope or plastic rope threaded through. The film strips should be spread from bottom to top in a way that allows the upper film strip to overlap the lower one with a lap width of about 30 cm. After that, the front roof should be pressed and tensioned by fixing the upper end of the tension ropes on the ridge purlins, and the lower end on the ground anchors or the pre-fixed expansion hooks.

6. Cold-proof Ditch

Cold-proof ditches are made at the front bottom corner or outside the rear wall in the south of the greenhouse, with a width of 30–40 cm, a depth of not less than that of the local frozen soil layer, and a length of slightly greater than that of the greenhouse. The ditches should be filled with thermal insulation materials such as weeds, straw, and rice hulls, covered with 10–15 cm thick soil, and then mulched to prevent rainwater from seeping in. It is also feasible to bury polystyrene boards with a thickness of 10 cm and a depth of not less than that of the local frozen soil layer to replace the cold-proof ditches.

7. Entrance and Exit

The entrance and exit, generally set on the side wall or rear wall and also on the front roof, should not only serve as easy access to the greenhouse but also should be conducive to thermal insulation. They are generally 1.5–2 m high, but some farmers make them only 60–80 cm high to achieve thermal insulation. To prevent cold air from leaking into the greenhouse, the entrance and exit are zigzag-shaped, with cotton curtains hanging inside and outside. A workroom should be built outside the greenhouse to store production materials or rest.

8. Thermal Insulation Coverings

The front roof of the greenhouse should be mostly covered with straw mats or thermal insulation quilts to reduce heat dissipation at night. The straw mat is generally 1.2–1.4 m wide, 1.5–1.8 m longer than the framework, and 4–5 cm thick. The two straw mats should overlap each other by about 20 cm and extend to cover 50 cm of the east and west side walls. Thermal insulation quilts have the advantages of a small heat transfer coefficient, good thermal insulation, moderate weight, easy to roll, good wind and water resistance, long service life, etc. It has developed a variety of insulation products, such as needled-punched felt thermal insulation quilts, composite thermal insulation quilts, acrylic staple fiber thermal insulation quilts, cotton felt thermal insulation quilts, and foam thermal insulation quilts.

To save labor, cut the curtain rolling time, and increase the light exposure of the greenhouse, the rolling of straw mats and thermal insulation quilts are mainly realized by greenhouse rolling machines. Currently, there are two main types of greenhouse rolling machines: fixed type and movable type. The fixed rolling machine is installed on the rear wall of the greenhouse. It rolls up the straw mats or thermal insulation quilts with a gear motor. The straw mats or thermal insulation quilts are then released under gravity along the slope of the greenhouse. The movable rolling machine is on the front roof and rolls the straw mats and thermal insulation quilts up and down as it moves upward and downward the roof.

Task 3 Control of Environmental Conditions in Facilities

I. Temperature Control

1. Thermal Insulation

To improve the heat retention capacity in the facility, the following methods can be used: Improve the airtightness of the facility by reducing the cross-flow of air and ventilation volume, increasing the layers of coverings and providing thermal insulation coverings with good insulation properties; increase the thermal insulation ratio, i.e. appropriately lower the height and reduce the heat dissipation area of the protection facility at night; increase the surface heat flow, i.e. increase the heat storage of the soil during the day and reduce the soil evaporation and crop evapotranspiration by increasing the light transmittance of the facility; cover the land surface; build cold-proof ditches to prevent the outward transmission of heat in the soil.

2. Heating

When solar radiation cannot meet the crop's temperature needs, the following manual heating measures that are efficient, energy-saving, and practical should be taken: hot water heating, steam heating, hot air heating, electric heating, and furnace heating. Hot water heating is characterized by good heat stability, stable and uniform room temperature, small fluctuation, safe and reliable supply of heat, and large heating loads, making it a commonly used heating mode in large and medium-sized greenhouses; steam heating and hot air heating work rapidly but show low-temperature stability; electric heating and furnace heating feature low construction and operation costs, but also low thermal efficiency.

3. Cooling

Strong solar radiation in summer leads to a temperature of up to 50°C in the facility, far exceeding the appropriate temperature for the growth and development of crops, which tends to cause high-temperature damage. According to the thermal equilibrium principle, the facility can be cooled by natural ventilation or forced ventilation, by shading with indoor or outdoor shade nets, by whitewashing the roof or by evaporation, such as through spraying water and using wet curtains. Ventilation is an important means of cooling. The principle of natural ventilation is ventilating from small vents to large ones, from middle vents to top ones and finally to bottom ones (these vents are closed in reverse order). The principle of forced ventilation is that the airflow should be kept away from the plants to reduce their impact on them and that many small vents are better than a few large ones. In winter, an exhaust fan is used to realize outward heat dissipation, thus preventing cold air from blowing to the plants and frostbiting them.

II. Light Control

1. Reasonable Facility Structure

The greenhouse should face south and extend from east to west. To strengthen daylighting, in addition to the modern multi-span greenhouses, single-span greenhouses with a proper span, height, and roof angle, as well as reasonable spacing from adjacent greenhouses should be selected where possible. Transparent covering materials that are dust-proof, drip-proof, and age-proof and with strong light transmittance should be selected. Currently, EVA film is preferred, followed by PE film and PVC film. Fine and solid framework materials should be selected where possible to improve the indoor lighting quantity and reduce the shading area of greenhouse structural materials.

2. Strengthening Facility Management

Regular cleaning and washing are required to maintain the high light transmittance of transparent covering materials on the roof. Under the premise of maintaining room temperature, opaque internal and external coverings (insulation curtains, straw mats, etc.) should be removed in the morning and added at night where possible so as to prolong the light exposure time and increase the light transmittance; VMPET reflection curtains should be hung in the greenhouse or the glass greenhouse roof should be whitewashed to increase the light intensity and the uniformity of light distribution.

3. Strengthening Cultivation Management

Select a proper planting density, pay attention to the row direction (generally the south-north direction is better), expand the row spacing, reduce the plant spacing, and remove the lateral branches and old leaves at the base of seedlings to increase the light transmittance of crops.

4. Timely Supplementing Light

The purpose of artificial light supplementing is to adjust the light period, requiring low-intensity light; or to promote photosynthesis and supplement the deficiency of natural light, requiring light intensity above the compensation point. There are three requirements for the electric light source used for this purpose: ①It must have a certain light intensity (the light intensity on the bed surface should be above the light compensation point and below the saturation point); ②The light intensity should be adjustable to a certain degree;③It must have a certain spectral energy distribution and can simulate the natural light with a continuous spectrum. Incandescent lamps are commonly used for cultivation under electric light, high-pressure gas discharge lamps for cultivation by supplementing light, and fluorescent lamps for both types of cultivation. The supplemental lighting is set below the inner insulation layer, and the reflective film is often used around the greenhouse to improve the light supplementing effect. Supplemental light intensity varies from crop to crop. Owing to the high cost of equipment, power consumption, and operation, light supplementing is only used for flowers with high economic value or seasonal production of seedlings.

5. Shading and Darkening as Required

Horticultural plants should be shaded or kept in darkness if they are short-day plants or need to oversummer or be blanched. In production, 25%–85% shade nets or aluminum foil composite materials are generally selected in accordance with the lighting condition. They are required to have certain light transmittance, high reflectivity, and low absorption rates. They should better be movable. A balance should be kept between temperature and light by opening and closing such coverings as appropriate. A glass greenhouse can also be sprayed with lime and other special reflective materials on the top to weaken the light intensity, but such materials must be washed off after summer. To keep the facility in darkness, such materials as black PE film and black braided fabrics can be used.

III. Humidity Control

1. Dehumidification

Dehumidification is mainly intended to prevent crops from getting wet and reduce air humidity to adjust plant physiological status and prevent diseases. There are active dehumidification and passive dehumidification based on whether power is supplied. Active dehumidification primarily depends on heating and ventilation (especially forced ventilation) to reduce indoor humidity. Passive dehumidification can be realized by natural ventilation (opening ventilation windows, removing the film, scratching seam, etc.); ground hardening or mulching; appropriate irrigation (drip irrigation, filtration irrigation, irrigation under the film, etc.); using of hygroscopic materials such as insulation curtains and non-woven fabrics; reducing evapotranspiration by plant adjustment.

2. Humidification

Under high temperatures and strong light in summer, excessively low air humidity is detrimental to crop growth and may cause plant withering or death in severe cases. For the cultivation of some flowers and vegetables requiring high humidity, the relative humidity must be raised if it is less than 40%. Common methods include atomizing and sprinkling with equipment such as electric atomizing humidifiers, air washers, centrifugal atomizers, and ultrasonic atomizers. The wet curtain cooling system can also increase relative humidity. In addition, lowering room temperature or light intensity can raise relative humidity or reduce transpiration intensity. Air humidity can also be increased with such measures as increasing the frequency and amount of watering and reducing ventilation.

IV. Control of the Gas Environment

1. Additional Application of CO_2 Gas Fertilizer

The proper concentration of supplemental CO_2 to be applied artificially depends on crop type, variety, and light intensity, as well as weather, season, and crop growth phase. From the perspective of photosynthesis, the CO_2 concentration close to the saturation point is the optimal concentration. In actual production, even under strong light, CO_2 concentration should not be increased above the saturation point. This is not only a waste of resources but also will result in the decrease of stoma opening and transpiration reduction, ultimately leading to CO_2 poisoning of plants, which is manifested as crop withering, chlorosis and defoliation. Therefore, 800–1,500 μL/L is usually the recommended concentration for most crops in production. CO_2 concentration should be below 1,300 μL/L on sunny days and 500–800 μL/L on cloudy days. However, it is not recommended to apply CO_2 gas fertilizer on rainy days. In general, fertilization should start 30 min after sunrise on sunny days and stop 30 min before ventilation to avoid waste. Applying for 2–3 h per day can effectively avoid the CO_2 "starvation" of the crops and increase the growth rate. CO_2 gas fertilizer may be applied through fermentation of organic fertilizer or crop straw, combustion of carbohydrates, and chemical reaction, or by using high-pressure CO_2 cylinders and dry ice.

2. Prevention of Harmful Gases

Gases harmful to crops can be prevented with the following measures: apply fully decomposed organic fertilizers, rather than unfermented fresh chicken manure or human excreta, or volatile fertilizers such as ammonia, ammonium bicarbonate, ammonium nitrate; reasonably apply chemical fertilizers; recommend fertilization in bands and holes and bury the bands and holes and water immediately after fertilization; select appropriate plastic products (ensure safety and non-toxicity) and timely ventilate the greenhouse to discharge harmful gases outside.

V. Soil Control

1. Additional Application of Organic Fertilizers and Organic Matter

Applying fully decomposed organic fertilizer can improve the physical and chemical properties of soil, make the soil loose and permeable, increase oxygen content, and help avoid the increase of salt concentration. Applying crop straw can improve the secondary salinization of soil. The graminaceous crop straw has a high C/N ratio, which contributes to the assimilation of nitrogen in the soil during the decomposition by microorganisms. Applying straw can also balance soil nutrients, increase the content of soil organic matter, promote soil microbial activities, reduce the number of pathogenic bacteria, and reduce diseases.

2. Formulated Fertilization

Formulated fertilization is to decide the appropriate amount and proportion of nitrogen, phosphorus, potash fertilizer, and trace elements, as well as the corresponding application techniques in accordance with the fertilization requirements of crops, soil fertilization characteristics, and fertilizer effects. Currently, the commonly used formulated fertilization techniques include soil nutrient balance and correction coefficient for soil available nutrients.

3. Reasonable Irrigation

The rising of soil moisture and evaporation through the surface are the main reasons for the accumulation of salt on the surface soil. Irrigation methods and quality are the main factors affecting soil water evaporation, so salt accumulation on the surface of the soil can be effectively prevented by reasonable irrigation. Drip irrigation and filtration irrigation are the most economical methods, which can prevent the subsoil salt from accumulating on the surface.

4. Crop Rotation, Soil Replacement, and Soilless Culture

Crop rotation and catch crop cultivation can alleviate secondary soil salinization and improve soil quality. Soil replacement is one of the effective measures to solve soil salinization, but it is rarely applied in practical production due to the high labor intensity and difficulty in obtaining satisfactory soil. In case of serious continuous cropping obstacles or rampant soil-borne diseases that cannot be solved with conventional methods, the soilless culture technique can be adopted.

5. Soil Disinfection

Normally, beneficial and harmful microorganisms in the soil maintain a certain balance, but this balance will be broken by continuous cropping, resulting in corresponding harm. In order to eliminate pathogenic bacteria or insect pests, soil can be disinfected with chemical agents (such as formaldehyde, sulfur powder, and nitrolime), steam, and solar energy.

6. Application of Microbial Fertilizer and Controlled-release Fertilizer

Applying microbial fertilizer can improve the physical and chemical properties of soil, enhance soil fertility, and increase the total amount and activity of beneficial microorganisms in the soil, effectively reducing continuous cropping obstacles. Controlled-release fertilizer is a new type of fertilizer that releases nutrients at a controlled rate and according to the nutrient demand of crops to achieve dynamic balance.

[Summary]

I. Key Points

Type of horticultural facilities.

II. Difficult Points

Environmental control techniques for various horticultural facilities.

[Skill Training]

Skill Training 6-1: Investigation on Types of Horticultural Facilities

I. Purposes and Requirements

(1) Through on-site investigation, measurement, and analysis of different horticultural facilities, combined with viewing of visual materials, deepen the understanding of the training content.

(2) Master the structural characteristics, construction methods, performance, existing problems, and cropping season of main horticultural facilities, as well as their status, role, and local application in horticultural crop production.

(3) Learn the structure survey method.

(4) Learn to identify and evaluate the rationality of horticultural facility components.

II. Tools and Equipment

1. Outdoor Investigation

Measuring tools such as measuring tapes, rulers, steel tapes, angular instruments (slope meter), and recording tools such as pencils.

2. Video Recordings and Equipment

Slides, videotapes, CDs, and other image data of different types and structures of horticultural facilities, as well as imaging equipment such as slide projectors, video players, and VCDs.

III. Contents and Methods

1. Field Investigation and Measurement

Divide the class into several groups, conduct field investigation by visiting the experimental bases of schools and the local production organizations in groups, and then sort out the measurement results and investigation data and prepare a report. The main requirements of the investigation are as follows.

(1) Investigate and identify the characteristics of several types of horticultural facilities, such as local greenhouses, and observe the site selection, facility orientations, and overall planning of the horticultural facilities. Analyze the similarities and differences in structure and performance, as well as the setting of energy conservation measures for various horticultural facilities.

(2) Measure and record the structure and specification, matching models and characteristics for various horticultural facilities.

① Measure and record the orientation and specification of cold or hot frames, seedbed layouts, and as the setting of cold frames.

② Measure and record the orientation, length, width, and height of small and medium-sized plastic arched sheds, as well as the types and specifications of framework materials and covering materials.

③ Measure and record the orientation, length, width, height, and arch spacing of plastic greenhouses (such as assembled steel pipe greenhouses and bamboo greenhouses), as well as the types and specifications of framework materials and covering materials.

④ Measure and record the orientation, length, span, and height of solar greenhouses and modern greenhouses, the angle of transparent roofs and rear roofs, the thickness and height of walls, the position and specification of doors, the types and specifications of construction materials and covering materials (transparent and opaque), as well as the types and allocation modes of supporting facilities and equipment.

⑤ Measure and record the structure, model, manufacturer, framework materials, covering materials (transparent and opaque), orientation, length, span, shoulder height, ridge height, spacing, and supporting facilities and equipment of large-scale modern greenhouses or multi-span greenhouses.

(3) Observe the equipment and performance of environmental control systems of modern horticultural facilities.

(4) Investigate and record the main cultivation seasons, varieties, annual utilization, and effective measures for energy conservation of various horticultural facilities in the local area.

2. Watch Image Data such as Videos, Slides, and Multimedia Data

Watch the image data of various horticultural facilities, such as simple ground facilities (simple and near-ground coverings), plastic film mulching, small horticultural facilities (small and medium-sized sheds), and large horticultural facilities (large sheds and greenhouses), to better understand their structural characteristics and application.

IV. Experiment Report

(1) Write an experiment report comparing various horticultural facilities by type, structure, performance, and application in the local area.

(2) Draw the structure diagram of the main types of facilities, indicate the name and dimension of each part, and point out the advantages and disadvantages, as well as improvement suggestions.

Skill Training 6-2: Structural Observation and Design of Plastic Arched Sheds

I. Purposes and Requirements

To preliminarily master the design methods and procedures of single-span plastic arched sheds, and draw their cross sections, stereograms, and plans through field investigation and measurement of their structures.

II. Materials and Tools

Measuring tapes, strings, steel tapes, special drawing tools, and paper.

III. Investigation and Design

1. Field Investigation and Observation

Conduct field investigation and observation at the experimental bases of schools or nearby production organizations by finishing the following tasks.

(1) Investigate the orientation of plastic arched sheds and the composition of framework materials.

(2) Accurately measure the span, length, and height of plastic arched sheds.

(3) Investigate the structural composition of plastic arched sheds. For arch sheds with bamboo-wood structure, investigate the setting of transverse and longitudinal upright columns, the column thickness (for greenhouses with steel-frame structure, investigate the structure of arch frame), the spacing of arch frames (bars), the setting of pull rods, the material of tension ropes, and the setting of ground anchors. For assembled steel pipe arch sheds, investigate the basic structure of arch sheds, as well as the types of accessories.

(4) Investigate and measure the setting and specifications of doors, as well as the types of covering materials.

(5) Investigate the auxiliary facilities of arched sheds, such as film reelers and multi-layer covering materials.

(6) Investigate the application season of plastic arched sheds in the local area, as well as the main crop species.

2. Structural Design of Plastic Arched Shed

(1) Based on the local natural and economic conditions, select appropriate plastic arched sheds, and determine their orientations and dimensions (length, span, height).

(2) Draw the outline of plastic arched sheds on coordinate paper.

(3) Determine the position and height of upright columns, the structure and spacing of arch frames (bars), and the setting of longitudinal pull rods.

(4) Determine the material of tension ropes and the position and specification of doors.

(5) Select suitable transparent covering materials, film reelers, and insulation curtains.

(6) Determine the depth of the plastic arched shed foundation and the depth and setting of the ground anchors.

IV. Experiment Report

(1) Based on field observation and investigation of plastic arched shed structures, summarize their advantages and disadvantages, carefully draw the plans and stereograms of the designed plastic arched sheds, and prepare design specifications and instructions for use.

(2) Specify the type, specification, and quantity of materials used in a designed plastic arched shed.

Skill Training 6-3: Application of CO_2 Fertilizer

I. Purposes and Requirements

Since arched sheds, solar greenhouses, multi-span greenhouses, and other facilities are relatively enclosed, CO_2 deficiency often occurs during plant growth. Generally, the CO_2 concentration is relatively high in the morning (500–600 ppm in the early morning) and evening. With the progress of photosynthesis, CO_2 in the facilities is absorbed by the leaves, resulting in a continuous drop of concentration. By around noon, the CO_2 concentration is very low (generally less than 100 ppm), which cannot meet the concentration requirement for the photosynthesis of vegetables in the facilities. The CO_2 in the facilities is often in short supply, seriously affecting photosynthesis, as well as the growth and development of crops. Therefore, applying supplemental CO_2 in protected fields has become an important means to reach high yield, high quality, and high efficiency in facility cultivation.

II. Planning

1. Materials and Tools

CO_2 gas testers, CO_2 cylinders, rulers, etc.

2. Investigation Plans

(1) Application Time of CO_2: CO_2 can be applied in the early stage of crop growth. Applying supplemental CO_2 in the seedling stage of fruit vegetables has a significant effect on shortening seedling age, promoting flower bud differentiation, and cultivating strong seedlings. The fruit setting and fruit swelling stages of fruit vegetables are the optimal periods for the additional application of CO_2. The time of CO_2 application varies in different seasons, generally 2 h after sunrise from November to next February, 1 h after sunrise from March to mid-April, and half an hour after sunrise from late April to May. The application in the seedling stage is generally 1.5 h after sunrise. CO_2 is generally applied once per day and it is necessary to close the shed 1.5–2 h upon completion of the application before ventilation.

Seedling culture stage: The concentration of CO_2 to be applied should be 0.5 mL/L at a time when the plants have 3–5 true leaves and can be gradually increased to 0.8 mL/L with the growth of plants. The application should be stopped 5–7 days before planting, and significant results can be obtained by continuous application for 18–20 days.

After planting: CO_2 should be applied continuously for 15–20 days after planting for 30–35 days, and the yield can be increased to the maximum. The application concentration of CO_2 should be 1 mL/L in sheds and 0.8–1 mL/L in greenhouses. CO_2 should be applied according to the required concentration on sunny days and reduced concentration on cloudy days. Application should be stopped on rainy and snowy days.

The daily application time should be subject to the daily change of CO_2 concentration in the facility. CO_2 should be applied half an hour after straw mats are removed in the morning, continuously applied for more than 2 h on sunny days, maintained at a high concentration, and discontinued 1 h before ventilation and in sunless weather or on rainy days.

(2) CO_2 Application Concentration: In closed greenhouses with sufficient light and vigorous crop growth, CO_2 deficiency often occurs. When the concentration is less than 80–100 mg/L, the normal growth of vegetables will be seriously restricted. The optimal application concentration of CO_2 is closely related to the crop type, growth stage, and weather conditions. Under suitable temperature, light, water, and fertilizer conditions, the photosynthetic rate of vegetable crops is generally fastest at 600–1,500 ppm CO_2 concentration. To be specific, the optimal CO_2 concentration is 1,000–1,500 ppm for fruit vegetables and 1,000 ppm for leafy vegetables.

III. Implementation

(1) Discuss in groups, consult data, and prepare CO_2 fertilization plans.

(2) Discuss the plans and revise them.

(3) Apply CO_2 as planned.

(4) Compare parameters.

(5) Summarize and evaluate the performance of trainees.

[Extension Tasks]

I. Review Exercises

(1) What are the measures for warming and insulation in horticultural facilities?

(2) What are the factors affecting the lighting in facilities?

(3) Please briefly describe the measures of humidity control in greenhouses from the perspectives of humidification and dehumidification.

II. Case Sharing

"Internet Plus" Intelligent Management System in Protected Agriculture

With the development of the Internet of Things (IoT), relevant products have been preliminarily applied in agriculture, including sensors in precision agriculture, intelligent expert management systems, remote monitoring systems, and agricultural product safety tracing systems. This contributes greatly to the promotion of agricultural techniques, the monitoring, prediction, prevention, and control of diseases and pests for agricultural plants, as well as to the scientific decision-making regarding agricultural production and operation.

Based on the actual needs and with the use of cutting-edge information technologies such as cloud computing, IoT, knowledge engineering, and 3S, an open, low-cost, intelligent management system with wide coverage has been developed to support the development of protected agriculture. With the use of IoT technologies, this system helps to create a relatively independent crop growth environment. It collects agricultural and ecological information about the facilities to realize intelligent management in combination with intelligent control and intelligent video surveillance technologies. Through the application of modern agricultural technologies and management methods, a sustainable agricultural development model aiming at realizing high yield, high efficiency, high quality, and ecological safety has been promoted to help realize intelligent, scientific, and intensive agricultural production and promote transformation and upgrading of modern agriculture, thus taking the development of protected agriculture to a new level.

Based on integrated communication technology, wireless network, and Zigbee technology, this system remotely acquires real-time data, including air temperature and humidity, soil moisture and temperature, CO_2 concentration, light intensity, and videos and images in the facilities through the sensor network, data acquisition module, and video surveillance module. The model analysis is conducted through intelligent control. Such equipment is being automatically controlled as wet curtains, spray and drip irrigation equipment, internal and external shades, roof windows, and side windows, warming and light supplementing equipment, and integrated water-nutrient control equipment to ensure optimal greenhouse environment for crop growth, and create conditions for achieving high quality, high yield, high efficiency and safety of crops. This system can also release real-time monitoring information and warning messages through information terminals such as mobile phones, PDAs, and computers.

It can be widely applied in protected agriculture, horticulture, and flower cultivation, or in places requiring a special environment, to timely present a scientific basis for realizing the healthy growth of ecological crops and timely adjusting cultivation, management, and other measures, and to realize regulatory automation. The establishment of the following systems: Establish a sound greenhouse intelligent control and acquisition system, an audio/video diagnosis and monitoring system, a peripheral video monitoring system, a regional environmental monitoring and display system, an integrated water-nutrient control system, and a comprehensive display service center, which are based on integrated communication technology, wireless network, and Zigbee technology. These systems can realize real-time display of data such as environmental factors, including air temperature and humidity, soil temperature and humidity, and CO_2 concentration, as well as video surveillance data; control the operation of shed rollers, ventilation, and irrigation equipment; send video information and intercom signals to the management center for remote video diagnosis, thus realizing remote intelligent management, training, and display. The systems can also realize real-time two-way audio/video communication, integrate videos and intercom signals, initiate real-time interaction between the management service center and an operator in any shed, and provide on-site technical guidance and training. The greenhouse data, video diagnosis data, and video surveillance data can be exchanged between the display service center and the comprehensive display center through high-speed network links to realize ICT-based management, centralized display, real-time preview, and view of video surveillance data of the park. Therefore, the interconnection and linkage of intelligent greenhouses can be realized, i.e., one person can manage fifty or even more greenhouses by controlling the greenhouses separately or together to meet the growth management requirements of different crops, greatly reduce manpower and material resources, create conditions for achieving high yield, high quality, high efficiency, eco-friendliness and safety of crops, and realize the intensive, networked and large-scale production management of the whole demonstration base. In addition, with the use of a computer-based automatic control system, the information can be collected, processed, analyzed, and released more quickly and accurately, remarkably reducing the workload. As the crop growing environment can be accurately controlled, the crop yield and quality can be significantly improved. The agricultural information provided for government departments serves as the basis for policymaking, guiding agricultural production and management, and reducing agricultural risks. In addition, considerable economic and social benefits have been generated, saving costs and increasing efficiency by over 10%, reducing labor by more than 10% on average, and improving yield and benefits by 10% on average.

Module 7 Techniques for Control of Lettuce Diseases and Pests

[Learning Objectives]

I. Knowledge

(1) Understand the diseases and pests of lettuce;

(2) Master the concept of infectious diseases and physiological diseases.

II. Skills

(1) Learn to retrieve relevant information;

(2) Be able to identify the diseases and pests based on relevant symptoms in the lettuce field and then take effective control measures.

[Preparation]

I. Required Resources

(1) A paper library and a periodicals reading room;

(2) A digital reading room and an electronic resource library;

(3) A multimedia classroom;

(4) Fields for observation.

II. Background Knowledge

(1) Master the basic methods of literature review;

(2) Understand the common diseases and pests of lettuce;

(3) Observe the specimens of diseases and pests of lettuce.

[Learning Tasks]

Task 1 Types and Control of Infectious Diseases

I. Damping–off

1. Symptom Identification

Lettuce damping-off is also known as seedling damping-off, small foot plague, which occurs after seedling emergence and before the unfolding of true leaves. Water-soaked spotting first occurs at the base of seedling stems on the ground; then the seedlings become yellowish-brown, dry out, and fall before the cotyledons wither. At the initial onset of the disease, often only a few plants are affected, withering during the day and restoring at night. Then 2–3 days later, the damping-off symptoms appear. In the case of high humidity in the seedbed, white flocculent hyphae are often found densely distributed on diseased seedlings or their adjacent bed surfaces.

2. Regularity of Occurrence

The damping-off pathogens overwinter in the soil with diseased residues in the form of oospores, spread to seedlings as zoospores through rainwater, irrigation water, and affected compost and agricultural tools, and invade from the stem base. Although the pathogens grow at an optimal temperature of about 15°C, the optimal ground temperature for the disease to break out is about 10°C, which is detrimental to seedling growth, but conducive to pathogen infection. Low temperature, high humidity of seedbeds, and insufficient sunshine are important factors that cause damping-off. Long-term rainy or snowy weather, and inadequate ventilation, light transmission, and thermal insulation of seedbeds result in excessively low seedbed temperature, and extremely rapid development of damping-off, which often leads to the death of a large number of seedlings. High water table and sowing density, untimely seedling thinning or transplantation, barren or viscous soil, and excessive watering are more likely to cause excessively humid and stuffy seedbeds, resulting in the prevalence of the disease. Removing the film untimely or improperly may easily lead to a significant change in the seedbed temperature and also easily induce damping-off. The plants are most susceptible to diseases, especially damping-off when the nutrients in cotyledon are all consumed, the root system is yet to become mature, and young stems have not been lignified.

3. Prevention and Control Methods

(1) Agricultural prevention and control.

Select the high and dry plots without succession cropping as seedbeds and adopt the rapid seedling raising or soilless culture method. Disinfect seeds, seedbeds, plug trays, and greenhouses before sowing. Strengthen the seedbed management: Pay attention to heat preservation during the low-temperature period by keeping the temperature within 20–30°C during the day and 15–18°C at night. Pay special attention to improving the ground temperature of seedbeds, reducing soil humidity and enhancing lighting. Apply plant ash in the early stage of the disease, and spray with 0.1%–0.2% potassium dihydrogen phosphate solution or 0.05%–0.1% calcium chloride solution at the seedling stage, so as to improve disease resistance.

(2) Chemical control.

Once a few diseased seedlings are found on the seedbed, remove them and spray with 500-fold-diluted 72% propamocarb aqueous solution, or 600–800-fold-diluted 25% metalaxyl wettable powder, or 500-fold-diluted 64% oxadixyl · mancozeb wettable powder, or 600–800-fold-diluted 75% chlorothalonil wettable powder, or 500-fold-diluted 70% zineb wettable powder, or 600-fold-diluted 50% carbendazim wettable powder, or 500–600-fold-diluted 58% metalaxyl · mancozeb wettable powder, or 500-fold-diluted 90% fosetyl-aluminium wettable powder, or 600-fold-diluted 30% hymexazol aqueous solution, twice to three times continuously, each at an interval of 7–8 days. Pay attention to spraying the seedling tender stems and the soil near affected seedlings. If a large number of dead seedlings are found, irrigate roots with 400-fold-diluted 72% propamocarb aqueous solution or 350-fold-diluted 55% metalaxyl wettable powder, twice to three times continuously, each at an interval of 6–7 days.

II. Downy Mildew

1. Symptom Identification

It may occur at all stages, from seedling emergence to harvest, and cause the most serious damage to adult plants. As it progresses from the base upwards, leaves are the most severely affected. In the early stage of the disease, light yellow subcircular to polygonal disease lesions are formed on the leaf surface. In the case of high air humidity, frosty mold layers will be formed on the leaf back, which can sometimes spread to the leaf surface. In the late stage, the disease lesions will cause the withering of a large area of plants that have turned yellowish-brown. In severe cases, all outer leaves will wither and die.

2. Regularity of Occurrence

The pathogens overwinter by harming seeds or autumn and winter crops in the form of hyphae, or overwinters on diseased residues in the form of oospores. Sporangia are produced in the following spring and are transmitted by airflow, watering, agricultural operations, and insects. They germinate at an optimal temperature of 6–10°C and infect plants at an optimal temperature of 15–17°C. The disease worsens in the case of a high planting density, high air humidity, or waterlogging in the field, premature or excessive watering after planting, long condensation time at night, or continuous cloudy days.

3. Prevention and Control Methods

(1) Agricultural prevention and control.

Select the disease-resistant varieties, such as Imperial and Great Lakes 366. Disinfect seeds before sowing: mix the seeds with 50% carbendazim wettable powder or 25% metalaxyl wettable powder with a weight of 0.3% that of the seeds. Completely remove the diseased residues after harvest and before planting; make high borders or use plastic film mulching; select a proper planting density and water the plants by dripping rather than flood irrigation; and timely drain the field after rain. Rotate the plants with Fabaceae, Liliaceous, and Solanaceae vegetables for 2–3 years.

(2) Chemical Control.

In the early stage of the disease, spray the back of the leaves at the stem base with 800–1,000-fold-diluted 69% enoyl · mancozeb wettable powder, or 600–800-fold-diluted 72% cymoxanil · mancozeb wettable powder, or 800–1,000-fold-diluted 66.8% propineb · iprovalicarb wettable powder. Preferably select dust agent or normal-temperature fumigant for control in protected fields: spray 1 kg of 5% chlorothalonil dust agent or fumigate with 0.5 kg of 45% chlorothalonil fumigant per mu.

III. Seedling Blight

1. Symptom Identification

It may occur throughout the seedling stage, generally in the middle and late phases of seedling raising. Early when the disease breaks out, oval lesions will occur at the stem base, and the diseased seedlings will wilt during the day and recover at night. Then, the disease lesions gradually depress, expand and encircle the stem. Finally, the stem base constricts and dries up, resulting in plant death. The disease develops more slowly than damping-off, and the affected seedlings do not collapse. In damp soil, pale brown cobweb mold will be found on the diseased parts, but it is not obvious. There is also no obvious white flocculent hyphae at the diseased parts, which sets the disease apart from damping-off.

2. Regularity of Occurrence

The pathogens overwinter in the soil in the form of mycelium or sclerotium and can be saprophytic in the soil for 2–3 years. Hyphae can directly invade the host and transmit through water flow or agricultural tools. The pathogens develop at an optimal temperature of 24°C. The disease is easily induced in the case of a high planting density, untimely seedling thinning, insufficient light, and excessively high temperature.

3. Prevention and Control Methods

(1) Agricultural prevention and control.

Select the high-lying plots with a low water table, good irrigation, and drainage conditions, and where solanaceous vegetables are not planted or preferably chives and leeks are planted as previous crops. Strengthen seedbed management by improving ground temperature, ventilating properly, reducing air humidity, and avoiding high air temperatures. Spray with 7,500–900-fold-diluted 0.1%–0.2% potassium dihydrogen phosphate solution or Zhibaosu in the seedling stage to enhance the disease resistance of plants; before sowing, mix an equivalent amount of 50% carbendazim wettable powder and 50% thiram wettable powder, then add 400-fold nutrient soil, and mix them well to effectively reduce the occurrence of diseases.

(2) Chemical control.

In the early stage of the disease, spray with 1,200-fold-diluted 20% tolclofos-methyl EC, or 500-fold-diluted 36% thiophanate-methyl flowable agent, or 1,500-fold-diluted 5% validamycin aqueous solution, or 450-fold-diluted 15% hymexazol aqueous solution, or 800-fold-diluted 30% carbendazim+thiram wettable powder. In the case of damping-off and seedling blight, spray with 800-fold-diluted 72.2% propamocarb aqueous solution + 800-fold-diluted 50% thiram wettable powder, based on a criterion of 2–3 kg of pesticide formulation per m^2, twice to three times continuously, each at an interval of 7–10 days depending on the condition.

IV. Powdery Mildew

1. Symptom Identification

It mainly damages the leaves and mostly spreads from the lower leaves to the upper ones. In the early stage of the disease, white powdery mildew spots are found on the leaf surface and back, and then a light grayish powdery mildew layer appears on the leaf surface. If conditions permit, these spots will expand to cover the entire leaf surface.

2. Regularity of Occurrence

The pathogens overwinter on diseased residues of lettuce or other hosts in the form of cleistothecia, or on lettuce plants in the form of hyphae. Once released from May to June next year, ascospores will spread by airflow and fall on the leaf surface, germinate infectious hyphae under suitable conditions, and invade the leaf epidermis. The spores germinate at a temperature of 10–30°C (20–25°C at the optimum). High humidity tends to cause a breakout of the disease, and high planting density and heavy application of nitrogen fertilizer will worsen the disease.

3. Prevention and Control Methods

(1) Agricultural prevention and control.

The following measures are recommended to prevent the disease: completely remove fallen and diseased leaves after harvest, and properly dispose of them in a centralized manner to reduce pathogens. Apply enough organic base fertilizer, additionally apply appropriate phosphatic fertilizer and potash fertilizer, strengthen field management, and enhance the disease resistance of plants.

(2) Chemical control.

In the early stage of the disease, spray with 8,000–10,000-fold-diluted 40% flusilazole emulsion, or 5,000-fold-diluted 30% triflumizole wettable powder, or 200-fold-diluted 2% pyrimidine nucleoside antibiotic aqueous solution, or 500-fold-diluted 5% wuyiencin aqueous solution, or 600-fold-diluted 40% sulfur · carbendazim flowable agent, or 4,000-fold-diluted 6% fenarimol wettable powder, twice to three times continuously, each at an interval of 7–10 days.

V. Gray Mold

1. Symptom Identification

It occurs mostly from the rhizome or lower leaves and then progresses from bottom to top, causing leaf wilting or death of the whole plant, and finally rot. It makes the rhizome ulcerated in the early stage and then rapidly progresses in all directions to rot the rhizome, causing gray mold layers on the diseased parts. In the case of high air humidity, the pathogen also infects the plant from its leaf margin or the leaf surface with water, forming round to oval spots. The leaf margin lesions are arc-shaped, initially water spots and gradually expand into yellowish-brown distinct ring rot, and form gray mold on the surface.

2. Regularity of Occurrence

The pathogen overwinters in diseased residues or soil as sclerotium or conidium. It is mainly transmitted by airflow or by undecomposed compost or watering. The pathogen grows at a temperature of 2–31°C (23°C at the optimum). The disease is easy to occur when the relative air humidity in the protected fields is greater than 90% and the temperature is 20°C, or when there are water droplets and wounds on the leaf surface, or when the plant shows weak growth.

3. Prevention and Control Methods

(1) Agricultural prevention and control.

Completely remove the diseased residues after harvest and adopt high-border cultivation, plastic film mulching, and drip irrigation technologies. Strengthen ventilation and lower the air humidity during the incidence period. Timely and carefully remove the diseased or infected plants and leaves identified: Put them in plastic bags, remove them from the greenhouse, and properly dispose of them. Rotate the lettuce with brassicaceous and goosefoot vegetables for 2–3 years.

(2) Chemical control.

Before planting, evenly spray the planting facilities with 500-fold-diluted 65% thiophanate-methyl·diethofencarb wettable powder, or 800-fold-diluted 40% pyrimethanil flowable agent, or 1,000-fold-diluted 45% thiabendazole flowable agent, or 400-fold-diluted 50% carbendazim wettable powder for surface sterilization. At the early stage of the disease, spray with 600-fold-diluted 65% thiophanate-methyl·diethofencarb wettable powder, or 1,000–1,500-fold-diluted 50% procymidone wettable powder, or 2,000-fold-diluted 40% dimethachlon wettable powder, or 1,000–1,500-fold-diluted 50% iprodione wettable powder, or 1,000-fold-diluted 50% vinclozolin wettable powder, twice or three times continuously, each at an interval of 7–10 days. Pay attention to spraying them alternately.

VI. *Sclerotinia* Disease

1. Symptom Identification

The disease may occur from the seedling stage to the adult-plant stage, mainly damaging the stem base. In the early stage, it gives the affected part yellowish-brown water spots. Then, it gradually expands to the whole stem, resulting in rot from leaf stalks to upwards, causing rotten stalks and leaves. Finally, it withers and kills plants. In the case of high humidity, flocculent and dense hyphae will form in the diseased parts, which later become black rat feces-like sclerotia. In dry weather, the diseased plants gradually wither and die, and the mold layer changes from white to grayish-green.

2. Regularity of Occurrence

The pathogen overwinters with diseased residues or in soil or mixed with seeds in the form of sclerotium. At the proper time, apothecium appears, and spores spread through air and irrigation. Spores invade from the senescent tissue or wound of the plant, namely the rhizome or basal leaves of the plant. The disease can be transmitted through direct contact of the affected leaves with adjacent healthy plants. The disease tends to break out when the temperature is about 20°C and the relative air humidity is higher than 85%. It will be significantly alleviated when the relative air humidity is lower than 70%, and will be aggravated in continuous cropping fields, low-lying fields, and fields with a high planting density and heavy application of nitrogen fertilizer.

3. Prevention and Control Methods

(1) Agricultural prevention and control.

Select disease-resistant varieties and disease-free seeds, and eliminate sclerotia mixed in seeds by sieving or seed soaking with 10% saline before sowing so as to cultivate strong seedlings at an appropriate age. Implement crop rotation for 3–4 years. Irrigate the field before soil preparation to eliminate sclerotia. Completely remove the diseased residues after harvest, plow the field deep for 50–60 cm, or cover it with a mulching film that can block ultraviolet transmittance so that the sclerotia cannot germinate or spores cannot disperse, so as to reduce pathogens. After removing the diseased residue, apply 0.2–0.3 kg of quicklime, 0.26–0.4 kg of cut rice straw or wheat straw every mu, and then turn up the soil to form borders and cover with the mulching film. Finally, close the greenhouse for 7–15 days to allow the soil temperature to rise to 40°C, so as to kill harmful germs. Strengthen management, weed out frequently, pay attention to ventilation, humidity control, and appropriate fertilizer and potash fertilizer application, and do not apply too much nitrogen fertilizer.

(2) Chemical control.

At the early stage of the disease, spray with 600-fold-diluted 65% thiophanate-methyl · diethofencarb wettable powder, or 1,200-fold-diluted 40% dimethachlon wettable powder, or 800-fold-diluted 45% thiabendazole flowable agent, or 500-fold-diluted 50% carbendazim wettable powder, or 1,000-fold-diluted 20% tolclofos-methyl EC, or 800–1,000-fold-diluted 70% thiophanate-methyl wettable powder by focusing on the stem base, old leaves, and the field surface, three to four times continuously, each at an interval of 7–10 days. Pay attention to spraying them alternately.

VII. Ripe Rot

1. Symptom Identification

It mainly harms the leaves. In the early stage, it gives the outer leaves small round brown ulcerative spots, which gradually expand into round brown or nearly round lesions with dark-colored edges and light-colored middle parts. In the late stage, the lesions become black and sometimes rupture to form small holes. The leaf midrib develops brown to dark brown long lesions from the base, which progress upwards to cause rots and depressions, resulting in yellowish-brown rot of the surrounding tissue. In the case of high humidity, dirty white to pink mold is found on the diseased parts.

2. Regularity of Occurrence

The pathogen overwinters with diseased residues in the soil in the form of mycelium or with conidium attached to the seed surface. It is transmitted by rainwater splashing and can invade the epidermis directly or from wounds. Warm and rainy weather and humid air are conducive to the disease. Disease will aggravate especially in rainy weather, and under rough management.

3. Prevention and Control Methods

(1) Agricultural prevention and control.

Perform strict quarantine to prevent the entry of diseased seeds. Soak the seeds in warm water at a temperature of 50–52°C for 10–20 min, and then cool and dry them for sowing, or mix them with 25% prochloraz wettable powder weighing 0.3% that of seeds, or with 50% carbendazim wettable powder weighing 0.4% that of seeds. Plow the field deep for soil solarization, apply enough organic base fertilizer, and additionally apply phosphatic fertilizer and potash fertilizer. Strengthen management, and timely perform drainage after rain to avoid waterlogging in the field. Pay attention to cleaning the field after harvest; rotate the lettuce with cruciferous vegetables; sow at a proper time, and avoid hot and rainy seasons.

(2) Chemical control.

At the early stage of the disease, spray with 600–800-fold-diluted 25% bromothalonil wettable powder, or 1,000-fold-diluted 25% prochloraz wettable powder, or 1,000-fold-diluted 25% propiconazole EC, or 400-fold-diluted 40% sulfur·carbendazim flowable agent, or 600-fold-diluted 70% thiophanate-methyl wettable powder, or 800-fold-diluted 68.75% famoxadone·mancozeb water dispersible granule, twice to three times continuously, each at an interval of 7–10 days.

VIII. Blight

1. Symptom Identification

Lettuce blight is also known as wilt disease. Most plants are infected on one side. At the onset of the disease, only the leaves near the ground turn yellow. Later, diseased leaves appear successively from bottom to top, turning yellow and brown; the stem is depressed on one side from bottom to top but grows normally on the other side. The disease progresses slowly, lasting 15–30 days from onset to death. In the late stage, only a few leaves remain on the upper part of the diseased plant, or all leaves wither. The anatomy of the stem and petiole of the diseased plant reveals that the vascular bundle turns brown but there is no ooze of white bacterial abscesses when squeezed by hand, which is different from bacterial wilt. When air humidity increases, a pink mold layer grows outside the diseased stem.

2. Regularity of Occurrence

The pathogen mainly overwinters on the diseased plant residues in the soil in the form of mycelium and chlamydospore. If it enters the seeds, the hyphae can also incubate in the seeds for overwintering. It will directly affect the seedlings after the seeds carrying pathogens are sown, or invade from the host root teloblasts or plant wounds and spread in vascular bundles. The resultant conidia move with the ascending fluid and are distributed in various parts of the plant including stems, leaves, and fruits. They germinate, grow into hyphae, and accumulate to block the vessels, making it impossible to transport water upwards, thus causing water loss, browning, necrosis, and withering of the plant. The pathogen can also produce fusarium in the host, which is harmful to plant tissue, makes vascular bundles brown, and accelerates plant death. Sometimes the pathogen moves upwards along the vascular bundle on one side of the plant, resulting in damage on one side of the plant. The disease will be serious in the field with thin surface soil layers, hardened bottom soil layers, and poor water permeability, but mild in the field with loose soil layers, good permeability, high organic matter content, vigorous plant growth, and developed roots.

3. Prevention and Control Methods

(1) Agricultural prevention and control.

Select seeds from disease-free plants and disinfect them if they carry pathogens. Select soil not planted with tomatoes for more than 3 years to make the seedbed. Perform disinfection if old seedbeds are used. Select Xi'an Dahong, Sukang 5, Yukang 4, Shuangkang 4, Mansi, Qiangfeng, Florida MH, and other disease-resistant varieties for local races. Rotate lettuce with non-solanaceous vegetables for 3–4 years, preferably by paddy-upland rotation. Strengthen cultivation management, evenly apply fully decomposed organic fertilizer as the base fertilizer, and then based on formula fertilization, apply a proper amount of potash fertilizer additionally to improve the disease resistance of the plants. Cultivate on high ridges to prevent waterlogging in the field; timely remove the diseased plants and deeply bury or destroy them.

(2) Chemical control.

Irrigate the roots with 500-fold-diluted 50% carbendazim wettable powder, or 500-fold-diluted 36% thiophanate-methyl flowable agent, or 500–600-fold-diluted 50% thiram wettable powder or 50% carbendazim wettable powder, or 200-fold-diluted 12.5% synergistic carbendazim wettable powder, or 500–1,000-fold-diluted 50% benlat wettable powder, 100 mL for each plant, three to four times continuously, each at an interval of 7–10 days.

IX. Stem Rot

1. Symptom Identification

This disease is harmful to leaves and petioles and mainly develops from the petiole near the stem base to the top. Irregular brown necrotic lesions are observed at the early stage, and then the whole petiole is affected, with dark brown sap overflowing at the lesions, resulting in rotted leaves. In the case of low air humidity, the disease is local and manifested as brown depressed lesions; In the case of high air humidity, there will often be mycelia or brown sclerotia growing on the diseased parts. In severe cases, the entire leafy head will be affected by wet rot and erosion.

2. Regularity of Occurrence

The pathogen mainly overwinters in diseased plant residues or soil in the form of mycelia and can survive in the soil for 2–3 years. It is transmitted through irrigation or agricultural work with hyphae invading from stomata on the leaf surface. The disease will be serious if the field is waterlogged, under a daily average temperature of above 20°C, or in an environment with high humidity; the disease will be extremely serious in fields with excessively high cultivation density, poor light transmittance, or heavy application of nitrogen fertilizer.

3. Prevention and Control Methods

(1) Agricultural Prevention and Control.

Select disease-resistant varieties and mix the seeds with 40% thiram · amicarthiazol wettable powder weighing 0.2% that of seeds before sowing to reduce seedling infection. After harvest, thoroughly clean the fields, deeply plow the soil, and rotate lettuce with rice to reduce pathogens. Select high and dry plots and perform drainage after rain in a timely manner to avoid waterlogging in the field. Maintain a properly high planting density, keep ventilation and light transmission in the field, and additionally apply a proper amount of phosphatic fertilizer and potash fertilizer to enhance the mechanical resistance of the plant epidermis.

(2) Chemical control.

At the early stage of the disease, spray with 3,000–4,000-fold-diluted 72% streptomycin sulfate soluble powder, or 1200-fold-diluted 20% tolclofos-methyl

EC, or 800-fold-diluted 50% thiram wettable powder, or 1,500-fold-diluted 5% validamycin aqueous solution, or 600–700-fold-diluted 72% cymoxanil · mancozeb wettable powder, or 250–300-fold-diluted 40% fosetyl-aluminium wettable powder, twice to three times continuously, each at an interval of 7–10 days. In severe cases, spray once to twice with 600–700-fold-diluted 72% cymoxanil · mancozeb wettable powder, and then 1,000–1,500-fold-diluted 70% fosetyl-aluminium · mancozeb wettable powder to achieve good effects.

X. Leaf Blight

1. Symptom Identification

The disease mostly occurs in the middle and late stages of growth, mainly harming leaves and petioles and progressing from bottom to top. Leaf infection mostly starts from the margin with grayish-brown to dark-brown irregular or polygonal lesions formed, or on the leaf surface with small brown spots formed that expand into large irregular spots, resulting in leaf withering. In the case of high air humidity, the disease will progress rapidly, and grayish-black mold layers will be formed on the surface of the lesions.

2. Regularity of Occurrence

The pathogen overwinters on diseased residues in the form of mycelia and conidia and is mainly transmitted through airflow, rainwater, and agricultural work. A warm and humid environment is conducive to the onset of the disease; weak plant growth, fertilizer deficiency, and waterlogging in the field will cause serious disease.

3. Prevention and Control Methods

(1) Agricultural prevention and control.

Soak the seeds in hot water before sowing. Select well-drained plots for cultivation, apply sufficient organic fertilizers and a proper amount of phosphatic fertilizer and potash fertilizer additionally. Apply supplemental fertilizer and water the plants at the right time to prevent premature senescence and enhance the disease resistance of plants. Completely remove the diseased residues after harvest to reduce the source of pathogens. Implement crop rotation for more than 3 years.

(2) Chemical control.

At the early stage of the disease, spray with 1,200-fold-diluted 50% iprodione wettable powder, or 1,500-fold-diluted 50% vinclozolin wettable powder, or 200-fold-diluted 2% pyrimidine nucleoside antibiotic aqueous solution, or 400-fold-diluted 50% captan wettable powder, or 600-fold-diluted 70% mancozeb wettable powder, or 500-fold-diluted 75% chlorothalonil wettable powder, or 600-fold-diluted 70% thiophanate-methyl wettable powder, once to three times continuously, each at an interval of 10–15 days.

XI. Soft Rot

1. Symptom Identification

The disease occurs at the middle and late stages of lettuce growth or in the heading stage, mostly infecting wounds at the base of the plant. At the initial stage, the affected parts are infiltrative translucent. With the progression of the disease, they gradually expand into irregular water spots, filled with light grayish-brown viscous secretion, releasing a putrid odor. The disease rapidly develops from the base upwards and rots the leafy head. Sometimes, the pathogen also infects from the outer leaf margin and the top of the leafy head, resulting in rot.

2. Regularity of Occurrence

The pathogen mainly overwinters on diseased residues of plants, or in soils and fertilizers, or by continuously harming other vegetables. It invades through wounds or fissures through irrigation, fertilization, and insect transmission. The pathogen grows at a temperature of 4–39°C (25–30°C at the optimum). The disease will be serious in fields that are low-lying or muggy, or continuously cropped, or with improper fertilizer and water management, or massive pests, or when numerous wounds of plants are caused during agricultural work.

3. Prevention and Control Methods

(1) Agricultural prevention and control.

Select disease-resistant varieties. Avoid continuous cropping and succession cropping, and rotate lettuce with graminaceous crops for 2–3 years. Apply fully decomposed organic fertilizers, plow and prepare the soils as early as possible, and cultivate on ridges or high borders. Sow at a proper time, avoid high temperature and rainy season during the susceptible period, avoid waterlogging in the field, and remove diseased plants as early as possible.

(2) Chemical control.

At the early stage of the disease, spray with 800-fold-diluted 47% kasugamycin·copper oxychloride wettable powder, or 500-fold-diluted 50% copper hydroxide wettable powder, or 4,000–5,000-fold-diluted 72% streptomycin sulfate water-soluble powder, or 800-fold-diluted 30% copper oxychloride flowable agent, or 400-fold-diluted 30% copper dihydroxosulphate flowable agent, or 1,000–1,500-fold-diluted 70% oxolinic acid wettable powder, once to three times continuously, each at an interval of 7–10 days.

XII. Black Rot

1. Symptom Identification

Lettuce black rot, also known as rot disease or bacterial leaf spot, is a bacterial disease that mainly affects leaves. At the early stage, orange subcircular spots appear on leaves and gradually progress into subcircular or irregular orange to dark-brown necrotic lesions, or the leaf edge becomes orange and the middle dark-brown depression, making them easily ruptured and perforated at the late stage. In severe cases, the lesions on the leaves are densely distributed and interconnected into pieces, resulting in necrosis of the diseased leaves. There is no odor when the lesions rot, which is different from soft rot.

2. Regularity of Occurrence

The pathogen overwinters through the diseased residues or seeds, invades from the stomata and wounds of seedling leaves in the following year, and later forms a systemic infection. Long-distance transmission mainly depends on the seed-carrying disease germs. Field transmission is mainly through rainwater, insects, and fertilizers. The disease breaks out more easily under high temperatures and high humidity and becomes serious in low-lying plots where succession cropping is carried out or many pests are found. The pathogen grows at a temperature of 0–38°C (25°C at the optimum) and dies at a temperature of 52–53°C. The pathogen can withstand dryness and has a long incubation period.

3. Prevention and Control Methods

(1) Agricultural prevention and control.

Select disease-free seeds or mix the seeds with 47% kasugamycin·copper oxychloride wettable powder weighing 0.3% that of the seeds before sowing. Rotate lettuces with bulb vegetables and graminaceous crops for 2–3 years to control pests. Cultivate by making deep furrows and high borders, maintain a properly high planting density, strictly prohibit flood irrigation, and reduce the field humidity. Apply fully decomposed organic fertilizers, avoid heavy application of nitrogen fertilizers, and additionally apply phosphatic fertilizer and potash fertilizer.

(2) Chemical control.

At the early stage of the disease, spray with 400–600-fold-diluted 47% kasugamycin·copper oxychloride wettable powder, or 600–800-fold-diluted 58.3% copper hydroxide dry flowable agent, or 800-fold-diluted 25% bismerthiazol wettable powder, or 350-fold-diluted 30% cuaminosulfate aqueous solution, or 800-fold-diluted 30% copper oxychloride flowable agent, or 500-fold-diluted 50% copper succinate wettable powder, or 500-fold-diluted 70% copper succinate·fosetyl aluminum wettable powder, once to three times continuously, each at an interval 10–15 days.

XIII. Leaf Scorch

1. Symptom Identification

The disease mostly occurs from the outer or innermost leaf margin, making the plant yellow from there to the petiole, with brown necrotic lesions formed. Then, the leaf margin becomes scorched with the progression of the disease, the leaf color becomes lighter, and finally the whole plant withers.

2. Regularity of Occurrence

The pathogen overwinters with seeds or disease residues and grows at a temperature of 4–41°C (28–30°C at the optimum). In the case of too much water loss from leaves, the infection which starts from the leaf margin, resulting in lesions and necrosis of leaf tissue, and gradually progresses and spreads to other parts. The disease will occur when there is a severe water loss of leaves due to high temperatures and low humidity in the field, or when the water absorption of the root system is affected by dry soil, low soil temperature, weak root growth, or high soil salt concentration.

3. Prevention and Control Methods

(1) Agricultural prevention and control.

Keep the soil moist, prevent field temperatures from being too high or too low, and maintain the normal water metabolism of plants. Apply ferment bacterial fertilizer to improve the physical-chemical proprieties of soils, avoid excessive watering and fertilization, and maintain suitable soil water content and suitable salt solution concentration. Spray foliar fertilizer if necessary.

(2) Chemical control.

Spray with 500-fold-diluted 77% copper hydroxide wettable powder, or 500-fold-diluted 50% copper succinate wettable powder, or 500-fold-diluted 60% copper succinate · fosetyl aluminum wettable powder, or 4,000-fold-diluted 72% streptomycin sulfate water-soluble powder, or 300-fold-diluted 14% cuaminosulfate aqueous solution, or 1 ∶ 1 ∶ 200 Bordeaux mixture for prevention and control.

XIV. Necrosis of Leaf Margin

1. Symptom Identification

Necrosis of lettuce leaf margin, also known as bacterial spot disease or root rot, mainly harms leaves. The disease mostly occurs from the head when it infects the leaf margin or nearby. It is initially manifested as water spots, which then become 0.5–1.5 cm wide, dry and paper-like, brown to dark brown, irregularly shaped oily lesions at the leaf margin, and reddish-brown spots on other parts of the leaves. Some lesions are gradually softened in succession, and some plants rapidly dry or defoliate completely. Generally, only the heading leaves are affected by the disease, which develops slowly and affects growth, but will not cause rot. For some plants, black-to-green hard rot tissue may occur in the center of the stem, which expands to the root along the veins of the bottom leaves when the plant is weak, causing root rot.

2. Regularity of Occurrence

The pathogenic bacteria overwinter in the soil and are transmitted by soil or air. The disease breaks out easily under low temperatures and high humidity and becomes serious in early spring and late autumn. Cold and foggy weather contributes to the disease and high temperatures in summer result in low incidence and slow progression of the disease.

3. Prevention and Control Methods

(1) Agricultural prevention and control.

Rotate lettuce with other vegetables. Strengthen field management, adopt formula fertilization technology, avoid high-humidity soil conditions, cultivate on high borders or high ridges, use plastic film mulching, and reduce the contact between diseased plants and healthy ones.

(2) Chemical control.

At the early stage of the disease, spray with 1,000-fold-diluted 47% kasugamycin·copper oxychloride wettable powder, or 500-fold-diluted 50% copper succinate wettable powder, or 300-fold-diluted 14% cuaminosulfate aqueous solution, or 500-fold-diluted 77% copper hydroxide wettable particle powder, twice to three times continuously, each at an interval of 10 days, and stop the spraying 3 days before harvest.

XV. Brown Rot

1. Symptom Identification

It occurs occasionally during the pre-reproductive period, mainly in the middle and late stages. The pathogen infects the plant from its rhizome or basal petiole. It is initially manifested as yellowish-brown water spots, gradually spreads upwards, or expands from the outer leaves to the inner ones, and finally causes the whole plant to turn yellowish-brown to dark brown and rot. In the case of high air humidity, plants will suffer soft rot with gray to grayish-brown filamentous hyphae sparsely distributed on their rhizome and petiole base. In the case of low air humidity, plants will turn light brown and die bccause of atrophy.

2. Regularity of Occurrence

The pathogen overwinters in diseased residues and soil in the form of mycelia or sclerotia and can survive through saprophytic nutrition in the soil. It has a strong adaptability to the environment and can grow at a temperature of 13–42°C (24°C at the optimum). A warm and humid environment is conducive to the onset of the disease. The pathogen may directly infect the plant under suitable conditions, mainly through irrigation, fertilization, and field agricultural work.

3. Prevention and Control Methods

(1) Agricultural prevention and control.

Before sowing, mix the seeds with 40% thiram · amicarthiazol wettable powder or 50% carbendazim wettable powder weighing 0.4% that of the seeds. Apply fully decomposed organic fertilizers, sow at a suitable time, avoid high temperatures and rainy seasons, and take cooling measures and timely water to improve the field ecological environment and enhance the disease resistance of plants.

(2) Chemical control.

At the early stage of the disease, spray 800-fold-diluted 45% thiabendazole flowable agent, or 600-fold-diluted 70% thiophanate-methyl wettable powder, or 600-fold-diluted 80% mancozeb wettable powder, or 600-fold-diluted 2% kasugamycin aqueous solution, or 500-fold-diluted 40% dichlozoline wettable powder, by focusing on the base of the plant, twice to three times continuously, each at an interval of 7–15 days.

XVI. Virus Disease

1. Symptom Identification

Lettuce virus disease is mainly caused by the lettuce mosaic virus and cucumber mosaic virus, and some are caused by the dandelion yellow mosaic virus. It may occur throughout the development period and has a great impact on yield if it occurs in the early stage. After four true leaves emerge from a seedling, the disease breaks out and gives light green or yellowish-white mosaic or mottles on the leaves, making them shrunken and crooked. The leaves have bright veins sometimes and even irregularly shaped gray to brown necrotic spots in severe cases. An adult plant affected by the disease will be significantly dwarfed, with irregularly shaped and twisted leaves. In severe cases, the thin veins of the plant will become brown, and many brown necrotic spots will be found on the leaf surface; anhydrous will appear, with a loose heading or without heading. For seed-collecting plants, the disease mostly occurs after their vine extension, resulting in mosaic or light and dark green mottles on new leaves. The blades will shrink, and the veins turn brown or have brown necrotic spots, causing weak plant growth.

2. Regularity of Occurrence

The virus mainly comes from the crops carrying viruses in the adjacent fields, and seeds can also carry viruses directly. If virus-carrying seeds are sown, the disease can occur at the seedling stage. The virus is mainly transmitted by aphids, including *Myzus persicae* (Sulzer), *Lipaphis erysimi* (Kaltenbach), and *Aphis gossypii* Glover in the fields, among which *Myzus persicae* (Sulzer) has the highest transmission rate. In addition, the virus can also be transmitted through contact with sap or friction. Weather conditions are directly related to disease occurrence. High temperatures and droughts will cause serious diseases. The optimal conditions for the onset of the disease include a temperature of 20–25°C and relative air humidity below 80%. Generally, the disease progresses rapidly and becomes severe under a daily average temperature higher than 18°C, or with a lack of water and fertilizer, rough management, or weed infestation.

3. Prevention and Control Methods

(1) Agricultural prevention and control.

Select disease-resistant varieties: Lettuces with scattered and wrinkled leaves are more disease-resistant. Sow disease-free seeds at a suitable time; completely remove diseased residues and weeds after harvest and before planting; cultivate on high borders and high ridges; use plastic film mulching; maintain a relatively low planting density; water frequently with small flows; strictly prohibit flood irrigation; timely perform drainage after rain. In the case of high temperatures and drought, keep the border surface moist to lower the temperature. Timely control and eliminate pests such as aphids and thrips.

(2) Chemical control.

At the early stage of the disease, spray 600–800-fold-diluted 72% cymoxanil·mancozeb wettable powder, or 800–1,000-fold-diluted 69% enoyl·mancozeb wettable powder, or 600–800-fold-diluted 50% carbendazim wettable powder, or 100-fold-diluted 10% fatty acid mixture aqueous solution, or 500-fold-diluted 20% moroxydine hydrochloride·cupric acetate wettable powder, or 500-fold-diluted 5% bacterial toxin aqueous solution, by focusing on the back of the basal leaves, twice to three times continuously, each at an interval of 7–10 days. Stop spraying 7–10 days before harvest. Preferably select dust agent or normal-temperature fumigant for protected fields. To achieve better effects, spray 1 kg of 5% chlorothalonil dust agent or fumigate with 0.5 kg of 45% non-restricted chlorothalonil fumigant per mu.

Task 2 Types, Prevention, and Control of Physiological Diseases

I. Top Rot

1. Symptom Identification

The disease occurs mostly during the heading stage or after the harvest of head lettuces. In the field, it causes chlorotic necrosis of the outer leaf margin, and water loss from the mesophyll tissue of the diseased leaves, which are grayish-white to grayish-brown, look papery, and roll slightly outwards. With the progress of the disease, the necrotic spots enlarge, while similar symptoms appear on multiple leaves, resulting in wilting, necrosis, and shrinking of the entire leafy head, and thus loss of commercial value. Affected by mild symptoms, the outer leaves will be basically normal, but the inner leaves will suffer irregularly shaped necrosis on the margin and shrink and look papery.

2. Causes of Disease

This disease mainly caused by an insufficient supply of calcium due to high temperature at the vegetative phase, or physiological calcium deficiency resulting from an imbalance between supply and demand of fertilizer and water.

3. Prevention and Control Methods

Prevent and control in an all-round manner in accordance with local conditions. Select plots with better irrigation conditions; perform soil preparation and water at a suitable time, especially in hot and dry weather during the heading stage; apply a proper amount of supplemental fertilizer; pay attention to the combined application of nitrogen, phosphorus, and potash fertilizers, and avoid heavy application of nitrogen fertilizers. If symptoms have occurred, spray the innermost leaves several times from the early stage of heading with 20,000-fold-diluted 0.7% calcium chloride solution + naphthylacetic acid or apply special calcium chloride sustained-release granules.

II. Nutritional Disorders

1. Symptom Identification

Nutritional disorders may refer to undernutrition due to the lack of a certain element or overnutrition due to excessive nutrients. In the case of nutritional disorders, leaves will present strong regular symptoms from top to bottom or from bottom to top. Generally, symptoms are not focused, but mostly sporadic and vary with the type of elements lacking.

(1) Nitrogen deficiency may cause weak plant growth and light green leaf color. The outer leaves are first affected, with their differentiation inhibited and the number reduced, which finally affects the yield. In severe cases, the whole plant is chlorotic, with old leaves falling off, young leaves stopping growth, and heading delayed or totally inhibited for the head lettuces.

(2) Phosphorus deficiency may cause inferior growth, dark green and lusterless leaves with a reduced number, resulting in yield decline. All of these symptoms first occur on the outer leaves.

(3) Potassium deficiency may cause the shrinking of lower old leaves and withered leaf margins.

(4) Calcium deficiency may cause rotten growing points, difficulty in expanding new leaves, curled or withered leaf apexes and leaf margins, as well as top rot.

(5) Magnesium deficiency may cause uniform chlorosis or etiolation between veins and green veins. Generally, the reticular mosaic will be found on leaves with withered leaf margins sometimes. The symptoms start from old leaves. In severe cases, the leaves become stiff with uneven surfaces and dead brown mesophyll at the yellow spots. The whole plant is largely chlorotic and brownish-yellow and purple somewhere and grows slowly and looks short with brittle leaves that are hardly edible.

(6) Boron deficiency may cause necrosis of growing points, blocked growth and shrinking of new leaves, and withered leaf margins; lateral branching of the stems and water-spot lesions in the pith part that are transversely cracked to form cavities; browning and wet rot of nearby flesh will result in collapse and withering finally.

(7) Copper deficiency may cause weak growth and pale leaf color, which first occurs on the outer leaves.

(8) Manganese deficiency may cause chlorosis of the middle and upper leaf apexes and leaf margins.

(9) Iron deficiency may cause blocked chlorophyll synthesis, uniform chlorosis of leaves, and severe chlorosis of top leaves.

(10) Molybdenum deficiency may cause chlorosis of the leaves and white or brown necrotic spots on the leaves.

(11) Excessive boron may cause the shrinking of old leaves and brown necrotic spots.

(12) Excessive chlorine may cause withering of the leaf margins of head lettuce, grayish-yellow to brown innermost leaves, but no obvious influence on the outer leaves.

(13) Excessive manganese may cause yellowish-white streaks on the leaves, which are connected to form yellowish-white spots.

(14) Excessive zinc may cause pale green veins in new leaves.

(15) Excessive copper may cause blocked growth and development, with iron deficiency symptoms in the new leaves.

2. Causes of Disease

This disease is caused by various external environmental factors, including nutrient imbalance, high temperature, drought, waterlogging, freezing, strong light, pesticides, and environmental pollution.

3. Prevention and Control Methods

Apply sufficient base fertilizer, adopt formula fertilization, and spray corresponding foliar fertilizer. For magnesium deficiency, perform top dressing with 5 kg of magnesium sulfate per mu, or spray the leaf surface with 0.5% magnesium sulfate solution; for boron deficiency, apply 1–2 kg of borax as the base fertilizer per mu, and spray the leaf surface with 0.1% borax solution at the early and middle stages of growth; for iron deficiency, spray the leaf surface with 0.2%–0.5% ferrous sulfate solution.

Task 3 Types, Prevention, and Control of Insect Pests

I. Aphids

1. Symptom Identification

Aphids are insects that secrete sugar-rich and sticky liquid waste. They penetrate the plant tissue by stylet to absorb the sap of leaves and young stems, resulting in the yellowing, shrinking, downward curling, and poor growth of leaves. Meanwhile, they carry the virus on the stylet, which will be transmitted to the affected plants, as one of the main ways of virus disease transmission. In addition, aphids can also secrete honeydew to contaminate leaves and cause sooty blotch.

2. Regularity of Occurrence

The threshold temperature for aphid's development is 4.3°C and the optimum temperature is 24°C (a temperature higher than 28°C is unfavorable for development). Aphids breed rapidly, totaling more than ten generations in one year in North China. The number of generations gradually increases from the North to the South, with a significant portion of overlapping generations. Under appropriate temperature and humidity conditions in the protected fields, aphids may affect the whole growth period of the plants. When aphids migrate, their migration direction is affected by color in addition to light and wind, showing a tendency to yellow and a strong avoidance of silver-gray.

3. Prevention and Control Methods

(1) Agricultural prevention and control.

Clean the field, remove weeds on the edge of the field and dispose of them in a timely manner. Plant some host plants around lettuce fields that aphids do not like to eat to avoid them and prevent diseases caused by them. Cultivate strong aphid-free seedlings and transfer plants to the field right after they are applied with pesticides. Expel aphids with silver-gray film or induce them by inserting or hanging yellow sticky boards 30–60 cm higher than the seedlings. To make the yellow sticky boards, wooden boards are first colored yellow and then applied with a layer of viscous grease after being dried. A total of 30–35 yellow sticky boards are used per mu and grease is applied once every 7–10 days. Condition permitting, use the natural enemy of aphids—green lacewings to prey on them.

(2) Chemical control.

Apply 500 g of 22% DDVP fumigant per mu before planting, and fumigate in a closed facility at night. To kill aphids found, spray 1,000–1,500-fold-diluted 80% DDVP emulsifiable concentrate, or 1,500–2,000-fold-diluted 70% fenitrothion wettable powder, or 3,000–5,000-fold-diluted 2.5% lambda-cyhalothrin emulsifiable concentrate for prevention and control, twice to three times continuously and pay attention to doing this alternately.

II. *Liriomyza*

1. Symptom Identification

Both larvae and imagoes of *Liriomyza sativae* can damage plant leaves, but the former is mainly a bigger threat. Imagoes are active during the day and inhabit the back of leaves at night; female imagoes stab the leaves with ovipositors in flight to feed on sap and scatter eggs in it, forming needle-tip-sized subcircular holes on leaves. The holes are light green at the initial stage and turn white later. They are visible to the naked eye. Larvae feed on the frontal mesophyll of leaves and form curved or coiled trails that are first narrow and then wide. The trails generally do not cross and overlap, and they widen significantly at the end. This is one of the characteristics that differentiate *Liriomyza sativae* from other flies. There are alternately arranged black frass on both sides of the trails, and brown dry plaque areas on the old trails in the late stages. Generally, one larva leaves one trail and a full-grown larva can tunnel about 3 cm within one day. Larvae crawl out of the trails to pupate on leaves or in soil crevices once fully grown.

2. Regularity of Occurrence

Liriomyza sativae grows and develops at a temperature of 20–30°C. They overwinter on vegetable residues in the form of pupae and imagoes and cause harm all year round in greenhouses. The temperature has a great impact on the activity of *Liriomyza sativae*: Their imagoes show weak motility at low temperatures and enhanced motility at high temperatures. Female imagoes stab leaves to feed on sap and oviposit in it; then eggs hatch in 2–5 days and the larval stage lasts 4–7 days. Full-grown larvae crawl out of the trails to pupate on the leaf surface or fall with the wind on the soil to pupate; pupae fledge to become imagoes in 7–14 days. They cause more damage in dry years than in rainy years. The imagoes of *Liriomyza sativae* show phototaxis, which is significantly more harmful to sun leaves or plants in winter than to shaded leaves or plants.

3. Prevention and Control Methods

(1) Agricultural prevention and control.

In the case of a small number of *Liriomyza sativae* occurring in the facilities or open vegetable fields, regularly remove affected leaves and burn them in a centralized manner, which has a certain control effect; or clean the fields, and burn or deeply bury the residual fallen leaves on the soil surface in a centralized manner. Deeply plow the soil for more than 20 cm to bury the pupae of *Liriomyza sativae* on surface soil into the deep soil layers, so as to make their eclosion impossible to reduce the fly source. Set a 20–24-mesh fly net at the facility vent to prevent the entry of imagoes. Sct yellow sticky boards to trap and kill imagoes in the fields by taking advantage of the fact that imagoes of *Liriomyza sativae* show a strong tendency to the color yellow. This has significant effects. In addition, use sticky flypaper to trap and kill imagoes: Set 15 trapping sites per mu from the initial peak period to the final peak period of imagoes, with one piece of flypaper placed at each site and replaced once every 3–4 days.

(2) Chemical control.

During soil preparation, it is recommended to evenly apply 1.5–2 kg of 3% isazofos granules or 3–4 kg of 5% phoxim granules in the soil per mu to eliminate overwintering pupae.

Apply 300–400 g of 32% DDVP fumigant, or 200–250 g of 30% chlordimeform fumigant for fumigation per mu of the facility, twice to three times continuously, each at an interval of 5–7 days. For any affected leaves, spray pesticides timely when the trails are very small. Spray 3,000-fold-diluted 1.8% avermectin EC, or 3,000-fold-diluted 10% cypermethrin EC, or 2,000-fold-diluted 5% flufenoxuron EC, or 2,000–3,000-fold-diluted 2.5% lambda-cyhalothrin EC, or 1,000–1,500-fold-diluted 20% bamectin·monosultap ME, or 1,500-fold-diluted 98% cartap raw powder for prevention and control.

III. *Trialeurodes vaporariorum*

1. Symptom Identification

Trialeurodes vaporariorum is also known as the greenhouse whitefly. Imagoes or nymphs swarm on the back of young leaves to suck sap with mouthparts, resulting in yellowing and withering of the leaves, and even death. They may also secrete a large amount of honeydew to contaminate the leaves, resulting in serious sooty blotch. Imagoes may also transmit certain virus diseases, resulting in yield drop and damaging commodity value.

2. Regularity of Occurrence

In the northern region, *Trialeurodes vaporariorum* cannot overwinter outdoors, with a developmental threshold temperature of 7.2°C and an optimal survival temperature of 22–27°C. More than 10 generations can be found throughout the year in protected fields. Imagoes have a strong flying ability, and a strong tendency to yellow and tender leaves. They prefer to oviposit on the back of new young leaves of damaged plants.

3. Prevention and Control Methods

(1) Agricultural prevention and control.

Plant low-temperature-resistant vegetable crops that *Trialeurodes vaporariorum* dislikes, including celery, garland chrysanthemum, spinach, oilseed rape, and garlic bolt in greenhouses seriously affected by *Trialeurodes vaporariorum*. Pay attention to eradicating sources of insects in the facilities, such as by removing residual weeds, fumigating residual imagoes, planting insect-free seedlings in the greenhouses, setting a fly net at the greenhouse vent to prevent the entry of external insects, and artificially releasing *Encarsia formosa* Gahan to control *Trialeurodes vaporariorum* hazards. When each plant has 0.5 *Trialeurodes vaporariorum*, release *Encarsia formosa* Gahan based on the criteria of 1,000–3,000 per m^2 per time, three to four times continuously, each at an interval of 15 days. Avoid planting gourd, solanaceous and legume vegetables that *Trialeurodes vaporariorum* prefer around the facility. Remove the branches and leaves with flies when pruning, or trap and kill the flies with yellow sticky boards.

(2) Chemical control.

Apply 500 g of 22% DDVP fumigant or 500 g of 10% aphis-gossypu-glover-killing fumigant per mu for facility cultivation, twice to three times at an interval of 7–8 days. Spray with 2,000–3,000-fold-diluted 10% imidacloprid wettable powder, or 1,000–1,500 fold-diluted 2.5% buprofezin wettable powder, or 1,000-fold-diluted 25% chinomethionat EC, or 2,000-fold-diluted 10% chlorfenapyr EC; better effects will be achieved by fumigation before spraying.

Task 4 Eco-friendly Techniques for Prevention and Control of Lettuce Diseases and Pests

I. Environmental Control

1. Temperature Control

Adopt multiple-covering cultivation techniques in winter, such as mulching film + small shed + non-woven fabrics to achieve good cold-proof and thermal insulation effects, effectively increasing the temperature by 3–5°C. Take temporary warming measures if conditions permit based on the principle of ensuring no freezing; pay attention to the density of fog and mist in the shed and timely ventilate it. When necessary, temporarily cover the field with white mulching film or non-woven fabrics on the surface for cultivation in the open field. Perform thermal insulation with sunshade nets and straw. Adopt cooling measures such as using sunshade nets in summer.

2. Humidity Control

Pay attention to ventilation and humidity control. If the temperature is high at noon on a sunny day, open the doors at both ends of the greenhouse for ventilation, but make sure to "apron" the lower part of the door to prevent cold wind from harming the vegetables. Pay close attention to the temperature change in the greenhouse, and close the air inlet in time.

II. Physical Prevention and Control

1. Trapping and Killing with Yellow Sticky Boards

This technique is mainly used to trap and kill whiteflies, aphids and *Liriomyza sativae*. The sticky yellow boards should be kept hanging from the seedling stage of vegetables or after planting. They should be hung vertically parallel to the crop rows, so as to trap and kill most target pests, with little impact on agricultural work. The bottom edge of the yellow boards should be 5 cm below the top of the plant or level with the top of the plant leaves, with a recommended suspension density of 30–35 boards/mu. When 85% of a yellow board is full of pests, replace it with a new one, which usually happens once every 30 days.

2. Trapping and Killing with Solar Pest-Killing Lamp

Frequency-vibrancy pest-killing lamps represent a new technique for trapping and killing pests that utilize their phototaxis and wave-taxis. They feature strong trapping and killing effects, little impact on beneficial insects, and convenient operation. Pest-killing lamps should be used in large areas continuously, and the single-lamp control area recommended is about 15–25 mu when there is sufficient light, and 30–40 mu when there is little light. The recommended height for hanging lamps is generally

80–100 cm. The daily light-on time generally ranges from 19:00 to 5:00 the next day. Since it is the optimum time for the pest-killing lamps to trap pests from the evening to 22:00 every day, it is best to keep the pest-killing lamps on from 19:00 to 24:00. In special cases, the light-on time can also be adjusted appropriately.

The trap bags of pest-killing lamps must be washed once every 3 days, and if used in summer when the temperature is high, they should be washed every day to improve the trapping and killing effects. The high-pressure contact net residue and other debris of pest-killing lamps should be cleaned once a day.

III. Biological Prevention and Control

1. Prevention and Control with Sex Attractant

The trapping and killing technique with sex attractants works by taking advantage of the sexual psychology of pests. With this technique, male flies are attracted to the traps by the synthetic female pheromone released from the lure and are killed, thus reducing the number of pests. The timing for application should be determined and adjusted in accordance with the occurrence time of target pests. Generally, they are used when the population density is low in the early stage. Pay attention to using them continuously.

The traps can be hung on a bamboo pole or wooden stick and well-fixed with their height appropriately adjusted according to the target pests and cultivated crops. Hanging them too high or too low will affect the trapping and killing effects. For larger pests such as cotton leafworm and beet armyworm, the trap height should be generally 0.8–1.0 m.

Traps should be arranged slightly densely on the periphery of the target field to induce the pests in the field, and relatively sparsely in the central part of the target field to trap the pests that have entered the target area, so as to improve the control effects. The trap density should be considered comprehensively in accordance with pest species, population base, application cost and other factors. Generally, one trap with one lure should be set every 2–3 mu for cotton leafworms and beet armyworms. The sex attractants should be applied in a centralized manner where possible so as to achieve better effects.

2. Prevention and Control with Predatory Mite

Use predatory mites to prevent and control whiteflies. Release predatory mites [*Amblyseius cucumeris* (Oudemans)] when the number of whiteflies per 100 leaves is 2, or release predatory mites in the field after half a month of application of pesticides to reduce the population base, based on the criteria of 50 bags (2,000–2,500 mites per bag) per mu.

[Summary]

I. Key Points

Correctly distinguish the common diseases and pests of lettuce in the local area.

II. Difficult Points

Formulate the prevention and control measures based on the characteristics of common diseases and pests of lettuce in the local area.

[Skill Training]

Skill Training 7-1: Investigation of Lettuce Undernutrition

I. Purposes and Requirements

(1) Master skills for the cultivation of lettuce seedlings;

(2) Master the preparation of the nutrient solution;

(3) Master the lettuce transplanting;

(4) Master the observation of lettuce undernutrition symptoms and description;

(5) Master the technique for the investigation of local fertilizer application.

II. Planning

(1) Instruments.

Beakers, several 5 mL and 1 mL pipettes, one 1,000 mL measuring cylinder, seven ceramic culture bowls, precision (pH 5–6) or universal pH test strips, glass rods, rubber pipette bulbs, and cotton.

(2) Reagents.

Potassium nitrate, magnesium sulfate, potassium dihydrogen phosphate, potassium sulfate, sodium sulfate, sodium dihydrogen phosphate, sodium nitrate, calcium nitrate, calcium chloride, iron sulfate, boric acid, manganese chloride, copper sulfate, zinc sulfate, molybdic acid, hydrochloric acid, and sodium ethylene diamine tetraacetate.

III. Implementation

(1) Materials: Select high-vigor lettuce seeds as experimental materials.

(2) Seedling cultivation: Wrap the seeds in wet gauze and put them in an incubator for pre-germination, then raise seedlings on planting sponge cubes and initiate hydroponics when the lettuce grows to have 4–5 true leaves.

(3) Stock solution: Prepare macroelement and trace-element stock solutions.

(4) Prepare balanced nutrient solutions and element-deficient nutrient solutions (pH 5.5–5.8) with deionized water (Table 7-1).

Table 7-1 Preparation of Balanced Nutrient Solutions and Element-deficient Nutrient Solutions

Stock Solution	Amount of Stock Solution per 100 mL of Culture Solution (mL)						
	Balanced	N Deficiency	P Deficiency	K Deficiency	Ca Deficiency	Mg Deficiency	Fe Deficiency
$Ca(NO_3)_2$	0.5		0.5	0.5		0.5	0.5
KNO_3	0.5		0.5		0.5	0.5	0.5
$MgSO_4$	0.5	0.5	0.5	0.5	0.5		0.5
KH_2PO_4	0.5	0.5			0.5	0.5	0.5

continued

Stock Solution	Amount of Stock Solution per 100 mL of Culture Solution (mL)						
	Balanced	N Deficiency	P Deficiency	K Deficiency	Ca Deficiency	Mg Deficiency	Fe Deficiency
K_2SO_4		0.5	0.1				
$CaCl_2$		0.5					
NaH_2PO_4				0.5			
$NaNO_3$				0.5	0.5		
Na_2SO_4						0.5	
EDTA-Fe	0.5	0.5	0.5	0.5	0.5	0.5	
Trace Elements	0.1	0.1	0.1	0.1	0.1	0.1	0.1

(5) Take two 700 mL ceramic pots and fill them with 600 mL prepared balanced culture solution and element-deficient culture solution, respectively, and label them (both on the pot body and lid) with a date.

(6) Wrap a cotton string around the stem base of plants with uniform growth, and carefully run the cotton string through the round hole to fix the stem, allowing the entire root system to be immersed in the culture solution.

(7) Place the pots in a place with sufficient sunlight and proper temperature for 3–4 weeks (replace with another balanced nutrient solution and element-deficient nutrient solution each week) after installation.

(8) Observe once every 2 days after the experiment begins, and check the pH value of the nutrient solution with a precision pH test paper. If the pH value is higher than 6, adjust it to 5–6 with diluted hydrochloric acid.

(9) Aerate the culture solution regularly every day, or leave a certain gap between the lid and the solution to fully oxygenate the root system. Water lost due to transpiration should be continuously replenished during culture (supplementing with deionized water).

Note the symptoms of deficiency of essential elements and where the symptoms appear first. After apparent symptoms appear in the pot of each element-deficient culture solution, replace all element-deficient solutions with complete ones; observe again to see whether such symptoms disappear, and keep records (Table 7-2).

Table 7-2 Observation Records of Experimental Phenomena

Plant Material:	Culture Time:		Observation Time:
Treatment	Symptoms of Overground Part	Symptoms of Underground Part	Analysis of Causes
Balanced Solution			
N Deficiency			
P Deficiency			
K Deficiency			
Mg Deficiency			
Ca Deficiency			
Fe Deficiency			

Skill Training 7-2: Pesticide Formulations and Application Methods

I. Purposes and Requirements

(1) Master common types of pesticide formulations and their application methods;

(2) Master the calculation of pesticide dosage and the preparation method of pesticide solution;

(3) Understand new pesticide formulations and their usage.

II. Experiment Content

1. Commodity Packaging Information of Pesticides

Take representative pesticide varieties of insecticides, fungicides, and herbicides, such as DDVP, carbendazim, and glyphosate, as the objects of demonstration, read the textual information on the outer packaging of pesticide commodities carefully, and identify the types of information, including the trade name, the name and content of active ingredients, formulation, application, usage, precautions, and storage methods.

2. Identification of Pesticide Formulations

Learn about the following three types of pesticide formulations commonly used in agricultural production, observe the physical morphological characteristics of each pesticide formulation, and understand the corresponding application methods.

(1) Solid preparation: Wettable powders, granules, powders, water-dispersible granules, and fumigants.

(2) Liquid preparation: Emulsifiable concentrates, aqueous solutions, emulsions in water, microemulsions, microcapsules, and flowable agents.

(3) New formulation: Attractant including insect sex pheromones.

3. Demonstration of Common Application Methods in Field Production (field demonstration of chemical equipment and multimedia-assisted demonstration are combined)

(1) Mist spraying method: It is about using a mist sprayer to convert pesticide solution into droplets, which are dispersed and suspended in the air, and then dropped onto crops or other treatment objects. It is one of the important application methods to prevent and control agricultural and forestry pests and can also be used in public pest control and disinfection.

(2) Powder spraying method: It is about using a powder sprayer to convert pesticide powder into fine powder, which is dispersed and suspended in the air and then dropped onto crops or other treatment objects. It is also one of the important methods of pest prevention and control in agriculture and forestry.

(3) Mixing with soil and broadcasting: It is about mixing the diluted pesticide with a certain amount of soil and then spreading the pesticide-carrying soil evenly in the field. It is mainly for preventing and controlling underground pests and seed-borne diseases and pests.

(4) Drip application: It is a new application method that has been successfully developed in recent years and is dedicated to paddy crops, especially rice fields. At present, the pesticides used in this method are mainly Monosultap and Bisultap. They are not easy to be absorbed by the soil but can be easily absorbed by the rice crop through the root system, and then transferred upwards to the rice stem, achieving satisfactory results. However, not all pesticides can be applied with this method, such as pesticides with low hydrosolubility that cannot be absorbed by crops but are easy to be absorbed by field mud.

(5) Aerial application: It is a way for the wide-range application of fertilizers by flying helicopters or UAVs.

(6) Stalk injection: It is about directly injecting chemicals into crop stalk with injection equipment.

(7) Soil fumigation: It is about placing a fumigant in soil under airtight conditions and preventing pests with the volatilizing of the pesticide.

(8) Application method of new pesticide formulations: Take insect sex pheromone as an example. First, install the lure on the trap as per instructions, and then hang the trap in the area infested by pests. Note: The trap should be installed somewhere high and hung upwind to facilitate the dissemination of sex pheromones.

4. Calculation of Pesticide Application Amount

(1) Pesticide (product) concentration.

Pesticide concentration refers to the concentration of technical material (active ingredient) in the mass of the chemical package, which is usually expressed by the mass fraction concentration. For example, 50% carbendazim wettable powder means the product contains 50 g carbendazim technical material in the mass of 100 g preparation.

(2) Pesticide (product) application amount in the field.

The pesticide application amount in the field is a numerical range recommended by the manufacturer, and it is usually measured in the following two ways:

① In "g/mu", that is, grams of pesticide formulations used per mu;

② With the dilution multiples of the formulation, such as 200-fold-diluted solution and 1,000-fold-diluted solution, indicating the dilution times per unit mass of the formulation.

In addition, the pesticide application amount in the field is sometimes expressed by the amount of active ingredient in "g/hm^2".

(3) Calculation and preparation of application amount.

When preparing a pesticide solution for field application, the amount of pesticide and water required should be calculated based on the crop area (taking conventional sprays as an example). If the pesticide seed dressing agent, seed coating agent, or other formulations, the seed mass for dressing should be determined as per instructions for pesticide use. In calculating, determine the amount of pesticide required first. Generally, the application amount of pesticide solution is 50–100 kg per mu and can be adjusted appropriately based on crop type. For vegetable and grain crops, the amount is generally 50–60 kg and occasionally 40 kg. For fruit trees, the amount can be increased appropriately, such as 100 kg. The corresponding amount of pesticide solution can be calculated according to the actual application area, and then the amount of pesticide and water required can be determined based on the recommended amount of pesticide.

In addition, to ensure more proper and accurate preparation of pesticide solution in the laboratory, it is often necessary to calculate the concentration of active ingredients of the pesticide solution (usually in "mg/kg", indicating milligrams of active ingredient per kilogram of the pesticide solution). To ensure the concentration of active ingredients in the finally prepared solution meets relevant requirements, it is necessary to specify the recommended amount of the pesticide and the amount of the pesticide solution per mu and calculate the dosage of the pesticide according to the amount of the pesticide solution to be prepared.

(4) Water adding method.

Water may be directly added when only a single pesticide is used. However, when several pesticides are mixed, only one portion of water will be added and the concentration levels of such pesticides will be calculated based on the same portion of water. For example, 500-fold-diluted urea plus 1,000-fold-diluted thiophanate-methyl is prepared by adding 2 portions of urea plus 1 portion of thiophanate-methyl and 1,000 portions of water. In addition, the mother liquor should be prepared first by diluting the pesticide solution with a small amount of warm water. Then, more water should be added to fully dissolve it and reach the required concentration, improving its efficacy and mitigating pesticide harm.

III. Experiment Report

(1) List 5 pesticide solid preparations and liquid preparations commonly used in production, and briefly describe the scope of application, application methods, and advantages and disadvantages of each formulation.

(2) Select 3 solid or liquid preparations, prepare 1 kg pesticide solution with them according to their recommended amount, and calculate the concentration of the active ingredient of the pesticide solution (unit: mg/kg).

[Extension Tasks]

I. Review Exercises

(1) What are the main symptoms of lettuce soft rot? How many types of pathogens are there? How can integrated control measures be prepared according to the source of infection and incidence conditions of pathogenic bacteria?

(2) What are the main symptoms of lettuce leaf blight? How to identify? Design comprehensive prevention and control measures according to the rules of occurrence.

(3) How to diagnose lettuce blight? What are the prevention and control measures?

(4) What are the main measures to prevent and control ripe rot and bacterial blight of lettuce? What are the differences in prevention and control methods?

(5) How to identify the symptoms of greenhouse whitefly and *Bemisia tabaci* infestation in the greenhouse? What are their main habits? How to prevent and control them in production?

(6) What are the symptoms and prevention and control measures for powdery mildew, seedling blight, downy mildew, and damping-off of lettuce?

(7) What are the characteristics of aphid infestation? Why does it occur on a larger scale and cause more serious damage in drought years or seasons? How to prevent and control them in production?

(8) Why does the vegetable leaf miner (*Liriomyza sativae)* spread rapidly in China? How to identify its symptoms? What are the prevention and control measures?

II. Case Sharing

Occurrence and Control of Vegetable Diseases and Pests

1. Occurrence and Hazard of Vegetable Diseases and Pests

There are many kinds of vegetable diseases and pests that cause serious damage. With the gradual expansion of the vegetable cultivation area in the protected fields, the incidence and harm of vegetable diseases and pests show a trend of further aggravation. Due to the frequent continuous cropping of vegetables in the protected fields, soil salinization becomes common, and root and stem diseases such as root rot, verticillium wilt, and blight are observed.

2. Integrated Control Technology for Vegetable Diseases and Pests

Vegetables are a class of crops with short growth cycles, quick rotation, and simple cooking processes, making them more susceptible to pesticide contamination, which causes great damage to human health. Therefore, in the prevention and control of vegetable diseases and pests, more emphasis should be placed on the coordinated use of multiple prevention and control measures, especially agricultural and biological control measures, and the usage of chemical pesticides, especially highly toxic and highly-persistent pesticides, should be strictly controlled.

(1) Prevention and control in overwintering.

① Deep tillage and irrigation in winter. In autumn and winter after the harvest of vegetables, deeply till or irrigate the land in time to expose a variety of pests and pathogenic bacteria that will overwinter in the soil to the ground, thus disrupting their overwintering environment. In addition, take advantage of the cold climate in winter and utilize mechanical equipment to reduce overwintering insects and bacteria sources.

② Cleaning fields and weeding. Promptly remove stubbles and leaves from the field after the harvest of vegetables in autumn. Remove weeds in and around the field in early spring to eliminate the pests and pathogenic bacteria attached to them and reduce egg-laying hosts and food that may be infested by the pests and pathogens.

(2) Prevention and control during seeding.

① Selection of disease- and pest-resistant varieties. Cultivating vegetable varieties that are resistant to diseases and pests is the most effective measure to control the hazard of diseases and pests. Through the experimental demonstration of introducing new varieties, superior varieties suitable for various cultivation conditions at home and abroad are purposefully selected to enhance the stress resistance and disease resistance of crops and realize high yield and high efficiency.

② Reasonable planning of planting pattern. Avoiding continuous cropping of cruciferous vegetables in a small area, large-scale cultivation of a single variety can mitigate a variety of diseases and pests. Vegetables of the same family are affected by the same or similar diseases and pests. Reasonable rotation is beneficial to vegetable growth. It can also reduce the accumulation of pathogens of perennial diseases in the soil and the food sources of monophagous and oligophagous pests. For example, intercropped or interplanted cabbages and mint or tomatoes can prevent the oviposition of cabbage butterflies (*Pieris rapae*).

③ Appropriate adjustment of seeding time or selection of early-maturing varieties. Adjusting the seeding time of vegetables or selecting early-maturing varieties where possible can avoid the incidence peak of some pests or the migration of insect vectors, thus reducing the damage caused by diseases and pests. For example, reasonable late sowing of cruciferous vegetables can reduce the occurrence of virus diseases. When cultivating cabbage, the use of protected cultivation with early spring mulch film or low-tunnel film can prevent cabbage caterpillars and reduce the occurrence of root diseases. Reasonable early sowing of spring garlic can avoid the oviposition peak of garlic maggot imagos during the germination period, thereby reducing the damage.

④ Cultivation of strong disease- and pest-free seedlings. The cultivation of strong disease- and pest-free seedlings is one of the key measures for protected cultivation in winter, which can prevent the occurrence of a variety of diseases and insects. This is especially important for the prevention and control of pests that cannot overwinter under open cultivation conditions in the North, such as greenhouse whiteflies, broad mites (*Polyphagotarsonemus latus),* and *Liriomyza sativae*, as well as zucchini virus disease and ginger rot. Select a leeward and sunward plot as the seedbed, and remove the residual roots, leaves, and weeds in the seedbed in time before raising seedlings. Use nursery pots or plug trays for seedling raising, prepare nutrient soil with disease-free seedbed soil, fully decomposed organic fertilizer, and a small amount of inorganic fertilizer, and disinfect the soil before sowing. Strengthen seedbed management after the emergence of seedlings and select high-quality, age-appropriate seedlings for planting.

⑤ Chemical treatment of seed and soil. Seeds or soil can be treated with high-efficiency and low-toxicity pesticides during sowing to prevent and control various underground pests and soil-borne diseases such as root-knot nematode disease, blight, sheath blight at the seedling stage, and damping-off. For various bacterial diseases, treating affected soil with pesticides is an important measure to mitigate their spread. To control bacterial diseases, formaldehyde can be used for seed disinfection; to prevent fungal diseases, fungicides such as thiram or carbendazim can be used for seed dressing. Pesticide dosage and time for seed disinfection depend on the vegetable varieties.

3. Prevention and Control in the Vegetative Phase

(1) Agricultural prevention and control.

The main purpose is to strengthen cultivation management and reasonable watering and fertilization. Fertilization should be based on proper fertilizer formulas and the fertilization needs and characteristics of various types of vegetables. The type of supplemental fertilizers (such as foliar fertilizers) should be properly selected to enhance plant growth vigor and resistance to disease, as well as reduce disease occurrence. Watering should be properly controlled, i.e. spraying a small amount of water frequently and avoiding flood irrigation, from the time of planting to the early stage of growth. Diseased and old residual leaves, flowers, and other diseased residues should be removed in time at noon on sunny days to reduce the damage caused by diseases and pests in the production process.

(2) Biological prevention and control.

The main way of biological prevention and control is to vigorously advocate the use of biologicals and protect the natural enemies of pests. For example, microbial insecticides such as *Bacillus thuringiensis* emulsion and insecticidal bacteria "qingchongjun" can be used to control lepidopteran pests such as cabbage butterflies and cabbage moths; antibiotics such as avermectin to control *Liriomyza sativae*, mites, greenhouse whiteflies, and lepidopteran larvae; streptomycin sulfate and oxytetracycline to control bacterial diseases such as cabbage soft rot and angular leaf spot; polyoxins and antimycoin to control downy mildew and powdery mildew; NPV formulations to control virus diseases; Liuyangmycin to control spider mites on legumes, melons, eggplants, etc. In terms of protection and utilization of natural enemies, predatory natural enemies such as ladybugs, grass predatory mites, and predatory flower bugs *Orius similis* Zheng are used to control pests. If conditions permit, *Trichogramma evanescens* Westwood can be released in the vegetable field to control cabbage moths, and *Encarsia formosa* can be released to control greenhouse whiteflies. Promoting the application of plant-derived pesticides, such as spraying leaching solution of mugwort leaves, cucumber leaves, and bitter gourd leaves, can prevent and control various vegetable pests; spraying leaching solution of pepper and tobacco can effectively prevent and control pests such as aphids, whiteflies, and red spiders.

(3) Physical and ecological control.

Installation of fly nets can effectively prevent the invasion of various pests. Aphids can be evicted by spreading silver-gray film or hanging silver-gray film strips in the field as aphids favor yellow and hate silver-gray. Aphids, glasshouse whiteflies, vegetable leaf miners, and the like can be lured and killed using yellow sticky boards. Some phototropic pests can be lured and killed by light. Cabbage moths, cotton leafworms, and the like can be lured and killed using sweet and sour liquid and sex attractants. The following key technologies can be adopted for the production of vegetables in the facility: greenhouse drip irrigation, double film insulation, high-border grafting and cultivation, full coverage with mulch film, high-temperature disinfection, and warming and insulation with fireplaces or firewalls in winter. The facility environment should be improved where possible to create environmental conditions suitable for vegetable plant growth and unfavorable for the occurrence of pests and diseases, thus minimizing the damage caused by diseases and pests. The temperature and humidity of the microenvironment of the protected area should be artificially regulated to inhibit the growth of certain pathogens. For example, during the fallow period of greenhouse vegetables in summer and autumn, the land should be deeply plowed, leveled, thoroughly watered, and covered with mulch film. The shed should be covered with mulch film and sealed for 7–10 days when it is sunny to raise the maximum temperature in the greenhouse up to 50–70°C, thus effectively killing the pathogenic bacteria and pests in the soil. In the production of greenhouse vegetables in winter, measures such as double-film covering, thickening of the straw curtain, increasing heat and light, excavation of cold-proof ditches in front of the greenhouse, hanging of reflective film on the rear wall, and application of EVA anti-drop film, can elevate greenhouse temperature, improve the light conditions, and effectively reduce the occurrence of freeze injury and various diseases.

(4) Chemical control.

The use of synthetic chemical pesticides on vegetables should be avoided. Where pesticides have to be used, high-efficiency, low-toxicity, and little-residue pesticide varieties should be selected in accordance with the standards for the safe and rational use of pesticides commonly used in vegetable fields. Attention should be paid to the proper use of pesticides, including the proper concentration, dosage, frequency of use, and re-entry interval. Pesticides should be used alternately to prevent resistance to them. When multiple diseases and insect pests attack concurrently, pesticides may be combined in use to prevent and control multiple diseases and insect pests together. The pesticide formulation and application method should be selected flexibly according to the weather change. On rainy days, smoke or dust agents should be used for prevention and control, which can effectively lower the humidity in the facility and reduce the damage caused by diseases and pests. The pesticide residue degradation technology should be promoted, i.e. efficient degrading bacteria should be used to eliminate pesticide residues in soil, water bodies, vegetable plants, and their products, generally 2 days after the pesticide is sprayed. Alternatively, 400-fold-diluted Bamulan, a non-toxic protective agent, can be sprayed before pesticide application to form a polymeric lipid film on the foliage, thus preventing pesticide contamination.

Module 8　Lettuce Breeding

[Learning Objectives]

I. Knowledge

(1) Understand the brief history and objectives of breeding lettuce;

(2) Understand the current status and future trends of lettuce breeding.

II. Skills

(1) Master breeding methods and techniques of conventional varieties;

(2) Be proficient in seed testing.

[Preparation]

I. Required Resources

(1) A paper library and a periodicals reading room;

(2) A digital reading room and an electronic resource library;

(3) A multimedia classroom.

II. Background Knowledge

(1) Understand the basic methods of literature review;

(2) Understand the relevant knowledge of plant breeding;

(3) Visit the lettuce breeding demonstration base.

[Learning Tasks]

Task 1 Breeding Overview

I. Brief History of Breeding

Lettuce originated along the Mediterranean coast. It is thought to have spread from Egypt to Europe via the Middle East and then to America by the Spanish in 1494. It was introduced into China in about the 5th century, and purple lettuce has been documented since the 11th century. Through long-term cultivation in China, lettuce has evolved into a stem-type one: asparagus lettuce (*Lactuca sativa* var. *angustata*). Much research has been done in developed countries such as the United States, the Netherlands, the United Kingdom, and France. As early as 1929, Jager of the United States developed a series of Great Lake varieties with large and heavy leafy heads, suitable for low-temperature storage and transportation. Since then, American breeders have developed a series of Imperial varieties that outperform Great Lake in terms of color, size, weight, bolting resistance, and disease resistance. From the 1980s to the present, the United States and Europe have successively developed the high-yield Vanguard series and Salinas series, as well as Target, Nancy, Clarion, Ardrade, and other head varieties. Maxon Smith and Ritehie in the United Kingdom developed six varieties suitable for low-temperature growth, of which Ambassador features light green leaves, early maturity, high yield, and low light tolerance, and is the most suitable for greenhouse cultivation. Baroet features light green leaves and extremely early maturity and it is suitable for early maturity cultivation without warming facilities or in an open field. Japan has also developed some crisp-leaf head varieties and soft-leaf head varieties that are resistant to low temperatures and mature extremely early. In addition, varieties that are resistant to high temperatures and bolting are also continuously bred. The research on the breeding of leaf lettuce in China mainly focuses on the introduction. Chen Yunqi et al. tested and compared eight leaf lettuce varieties introduced in Shandong Province. They compared and evaluated the development period, net vegetable rate, and vitamin C content of crisp-leaf head lettuce and soft-leaf one. Zheng Xianghong et al. piloted the planting of 128 leaf lettuce varieties introduced from the United States, the Netherlands, Japan, and other countries. They selected early-maturing, mid-maturing, late-maturing, disease-resistant, and high-yield crisp-leaf head varieties. Through testing and comparison, they chose Carona, Kobin, and Salinas varieties suitable for different cultivation conditions. Fan Shuangxi et al. tested and compared 205 leaf lettuce varieties introduced in Beijing, and bred four high-temperature-resistant and bolting-resistant varieties and four superior low-temperature-resistant varieties. In terms of asparagus lettuce breeding, varieties such as chicken leg asparagus lettuce, Jigua asparagus lettuce, and Huaye asparagus lettuce selected by Chinese breeders have their own characteristics and are widely cultivated throughout the country.

II. Current Situation and Development Trend of Breeding

1. Current Situation of Breeding

In recent years, some research institutes in China have carried out studies on lettuce germplasm resources. At present, China has 740 lettuce germplasm resources, including 532 stem lettuce and 208 leaf lettuce. Among them, 680 have been listed in the *Catalogue of Vegetable Varieties in China* (1992–1998), including 502 stem lettuce and 178 leaf lettuce. The rich germplasm resources of lettuce lay a solid foundation for genetic breeding. Wang Yanan analyzed the genetic relationship and genetic diversity of 47 purple-leaf lettuce germplasms with a TRAP marker. A total of 430 polymorphic bands were amplified from 20 primer pairs, with the percentage of polymorphic bands (PPB) being 67.08%. The genetic similarity coefficients of 47 purple-leaf lettuce ranged from 0.71 to 0.99, and the level of genetic diversity was low. The results of cluster analysis showed that germplasms with similar leaf morphology basically clustered together and were closely related. Stoffel et al. established a gene chip containing 6,500,000 characteristic data. It can be used to carry out polymorphism evaluation of a large number of leaf lettuce variety resources. In particular, it can identify many polymorphic sites between a similar variety of materials.

Lettuce has no obvious heterosis. Even if it has certain heterosis, its small floral organ makes hybrid and breeding difficult. Therefore, most of the varieties promoted in production are conventional varieties. Breeding efforts for leaf lettuce are mostly made in Europe and the United States. A large number of different special-purpose leaf lettuce varieties have been developed in Europe and the United States through conventional sexual crossbreeding, and all the leaf lettuce cultivated in China are of superior varieties selected and bred through the introduction of varieties from abroad. Stem lettuce is unique to China, and the stem lettuce that cultivated abroad on a small scale is also introduced from China. There are many local varieties of stem lettuce, and the current breeding is mainly based on investigation, introduction and selective breeding, that is, screening superior varieties from local varieties or selecting superior lines from the natural variation of local varieties to develop new ones. For example, 128 superior local varieties have been collected from local vegetable farmers by the Vegetable Research Institute of Guanghan, Sichuan since 1986, and 18 new high-temperature- and low-temperature-resistant varieties suitable for cultivation under different climatic conditions such as special high-temperature-resistant Large Erbaipi, special high-temperature-resistant Large Baijianye, and special high-temperature-resistant Large Authentic Baierpi were bred in the winter of 1994 through field observation and screening through studies. Mianyang Kexing Vegetable Development Co., Ltd. of Sichuan Province bred new special asparagus lettuce varieties Kexing No. 1 (round leaf) and Kexing No. 4 (sharp leaf) that are low-temperature-resistant and suitable for overwintering cultivation in areas with mild winter climate. Seedbanks in Yong'an and Sanming of Fujian Province bred varieties such as Feiqiao Lettuce No. 1.

In recent years, with the rapid development of biotechnology, traditional lettuce breeding techniques have evolved into modern molecular techniques, and certain progress has been made in some aspects. By optimizing all aspects of lettuce tissue culture to regenerate plants in vitro, high-frequency lettuce in vitro regeneration system has been established, which lays a foundation for the screening of lettuce mutants and transgenic research. Some researchers have used these high-frequency regeneration systems to carry out studies on lettuce genetically engineered vaccines, such as the study on the production of hepatitis B vaccines with lettuce as the bioreactor.

2. Development Trend

(1) Strengthening research on the introduction of germplasm resources.

There are many types of lettuce. Some cultivated in production lack resources and are closely related, with a narrow genetic background. Therefore, efforts should be ramped up to collect excellent resources from abroad, carry out the innovation of germplasm resources, make full use of the beneficial genes of wild and closely related species, and breed varieties in combination with distant hybridization to meet the current needs in leaf lettuce production.

(2) Breeding for resistance.

Considering the influence of climate change and varied pathogenic factors, the breeding of disease-resistant lettuce is still the main goal of breeders in various countries. In terms of breeding means, on the one hand, the existing resistance sources should be used to continue breeding disease-resistant and insect-resistant varieties; on the other hand, exogenous resistance genes should be introduced to expand and enrich resistance sources, and special varieties in areas with vertical resistance and widely applicable varieties with horizontal resistance should be bred to effectively control pests and diseases in the fields. Breeding of high-temperature-resistant and bolting-resistant varieties is also an important goal, especially for varieties for open cultivation.

(3) Breeding of special varieties.

According to the various types and production environments, different special-purpose varieties must be bred, such as varieties that stay fresh longer during storage and transportation, special varieties resistant to low temperature and low light for protected culture, open-field oversummering varieties resistant to high temperature, high humidity and bolting, varieties suitable for mechanized harvesting with consistent leafy head maturity period, size, shape and firmness, as well as varieties suitable for cultivation in the soil in different areas.

(4) Breeding of various varieties of leaf and stem lettuces with various maturity levels.

Varieties of different types and usages should be bred according to consumer demands, such as varieties adaptable to different seasons and early-, mid-, and late-maturing varieties.

Task 2 Breeding Objectives

I. Basic Requirements for Breeding Objectives of Different Types of Varieties

With the increasing popularity of lettuce in international and domestic markets, the breeding of lettuce has also received more and more attention from breeders in China and abroad. In general, the breeding objectives of lettuce include yield, quality, disease resistance, stress resistance and bolting resistance. Since lettuce includes two categories: leaf lettuce and stem lettuce, more specific breeding objectives are set for each category.

1. Leaf Lettuce

(1) Looseleaf lettuce. In terms of quality improvement, studies have shown that there is a highly significant positive correlation between the weight per plant, rosette leaf number, and the maximum leaf width of looseleaf lettuce. Therefore, to improve the yield of looseleaf lettuce, plants with many rosette leaves, wide leaves, and larger single leaves should be selected. In terms of resistance, attention should be paid to breeding varieties resistant to downy mildew and red spiders, as well as high and low temperatures. Since the edible organ is the leaf, bolting-resistant varieties should be selected.

(2) Head lettuce. In terms of quality improvement, as the head rate of head lettuce is significantly positively correlated with maximum leaf length, maximum leaf width, and head weight, and significantly negatively correlated with plant height and rosette leaf number, it is suitable to breed dwarf plants with large leaf

areas, heavy leafy heads, and few rosette leaves. At the same time, attention should be paid to the commodity quality of the product, and varieties with compact leafy heads, good shape, and good color should be bred. In terms of resistance, head lettuce is prone to many diseases, so the top priority should be given to breed varieties resistant to downy mildew, blight, virus disease, *Sclerotinia* disease, leaf blight, and other diseases. As with the looseleaf varieties, bolting-resistant varieties should be bred.

2. Stem Lettuce

The edible organ of stem lettuce is its fleshy young stem, so varieties with fleshy and stout stem, good color and good shape should be bred. Downy mildew, *Sclerotinia* disease, virus disease, and black spot of stem lettuce have been serious in recent years. High attention should be paid to the breeding of varieties resistant to such diseases. Also, bolting resistance is the main objective of breeding. Since there are many research reports on the breeding of leaf lettuce varieties, the following breeding objectives are explained using leaf lettuce as an example.

II. Breeding for Quality Improvement

In the early days, people mainly paid attention to such as traits thornlessness, reduced sap content, and leaf shape when selecting leaf lettuce varieties, which are the marking traits of varieties bred from the wild ones. As a result, most of the varieties that appeared in that period had sharp and narrow leaves, with no heads. With the progressing efforts of breeding, people have further developed the soft-leaf head type with a short vegetative phase, low demand for fertilizer and water, and low-temperature resistance. Later, the following varieties are further emphasized as the breeding objective: Crisp-leaf head varieties with large and heavy leafy heads are suitable for low-temperature storage and transportation as well as low-temperature sales, and they are resistant to moisture environments. The representative varieties were the Imperial series varieties. At present, the main objective of quality improvement is to utilize beneficial wild genes and closely related species through distant hybridization. Inferior traits of some varieties with poor texture, nutritional quality and taste are improved, taking into account other commercial traits, such as leaf color, head size and shape, and uniformity.

The development of Vanguard and its variants is an important sign of variety improvement. Thompson obtained Vanguard by distant hybridization and backcrossing with *L. virosa*. With dark green, smooth leaves, soft texture, flat veins, and developed root systems, this variety does not taste bitter at maturity and is less likely to rot with superior quality. Since then, a series of high-quality varieties such as Winterhaven, Moranguard, Vanguard 75, and Salinas have been developed. Vanguard varieties still dominate in the western United States.

Bolting resistance is also an important element of quality improvement. The Empire varieties developed in the United States in the 1980s are resistant to high temperatures and are not easy to bolt when planted in summer. The upright type Valcos cultivated by Leeper in 1991 is also a bolting-resistant variety.

III. Breeding for High Yield

The commercial part of leaf lettuce is the leafy head. There are various methods to assess its yield. In China, we mostly use the weight per unit area to calculate the yield of leafy heads of head lettuce. In terms of yield improvement, there are 4 factors that have a great impact, namely, disease control, leafy head volume improvement, uniformity, and net vegetable rate. These factors are very important for screening dominant individual plants in breeding.

IV. Breeding for Disease Resistance

Considering the influence of climate change and varied pathogenic factors, the breeding of disease-resistant leaf lettuce is still the main goal of breeders in various countries. In terms of breeding means, on the one hand, the existing resistance sources should be used to continue breeding disease-resistant and insect-resistant varieties; on the other hand, exogenous resistance genes should be introduced to expand and enrich resistance sources, and special varieties in areas with vertical resistance and widely applicable varieties with horizontal resistance should be bred to meet the demand for production and development and effectively control pests and diseases in fields.

1. Lettuce Downy Mildew (*Bremia lactucae* Regel)

Downy mildew is a sudden disease in lettuce production. Low temperature and humidity are conducive to its occurrence and prevalence, and it is more likely to occur during greenhouse cultivation. At present, at least eight physiological races of downy mildew have been found. As physiological races are easy to change and resistant varieties are likely to lose resistance in production, breeding for downy mildew resistance is difficult. Nevertheless, the resistance to downy mildew is a trait of race-specific type and is controlled by a dominant gene, and the disease-resistant genotype is easy to be identified. In the United States, the first generation of Imperial varieties are susceptible to downy mildew. The Valverde and Calmar developed in the second generation and later are also resistant to downy mildew, especially Calmar, together with leaf blight, and are moderately resistant to big-vein disease. They have rapidly replaced the Great Lake varieties in the western United States thanks to their strong resistance to downy mildew. The downy mildew resistance genes in these varieties are derived from the wild species *L. serriola*. Breeders of the Netherlands and France have also developed downy-mildew-resistant leaf lettuce varieties suitable for greenhouse cultivation, such as Clarion, Eortina, Ehifley, and Sitonia.

2. Big-vein Disease (Lettuce big-vein-associated virus)

Big-vein disease is common in lettuce production and can be transmitted through soil and grafting. Young plants infected with big-vein disease will show necrosis around the veins, and some plants will stop growing or their growth will be retarded due to blocked development, resulting in failure to form commercial leafy heads. Adult plants infected with the big-vein disease will show chlorosis of tissue around the veins, curling and hardening of external leaves, smaller leafy heads, and significantly reduced yield and net vegetable rate. An effective way to control the big-vein disease is to use disease-resistant varieties. In the United States, the Great Lake varieties are not resistant to big-vein disease, but Calmar, developed by Welch, is resistant to this disease and downy mildew, allowing it to be cultivated in large areas in production.

3. Virus Disease (Pathogens include lettuce mosaic virus, Dandelion yellow

mosaic virus, cucumber mosaic virus, and tomato aspermy virus)

The main virus disease is lettuce mosaic virus (LMV) disease. Virus diseases occur widely in all leaf lettuce production areas. The cotyledons and the first true leaf of seedlings infected with virus disease are severely deformed and typical lesions from light green to dark green will appear. Later, with the rolling of the leaf margin, the development of some plants is blocked, making the entire plant yellowish and unable to form leafy heads. The virus disease of leaf lettuce is mainly transmitted by virus-carrying seeds and insects, especially aphids. Seedlings developed from virus-carrying seeds are the source and spreading center of virus disease in the field. They will affect the whole field via insect transmission. Ryder of the United States reported the inheritance of the disease-resistant gene of leaf lettuce mosaic virus (LMV) in 1970 and concluded that the resistance gene is recessive monogenic inheritance. At present, most cultivated stem lettuce and leaf lettuce are not resistant to LMV. For example, the Great Lake and Imperial varieties are susceptible to virus disease. In contrast, Vanguard 75 and Florid 1974 are two newly developed varieties with strong resistance in the United States.

4. Blight (*Fusarium oxysporum* f. sp. l*actucae* Schlechtend.: Fr. f. sp.*lactucae* Matuo & Motohashi)

Blight is a perplexing disease in lettuce production, which occurs in many areas and causes serious harm, and has a tendency to gradually cause greater harm in a wider scope. In 2011, Chen Xiaoyu et al. first reported the occurrence of the blight of Beijing Tongzhou lettuce (head lettuce) and identified the pathogen of lettuce blight, which was found to be consistent with the *F. oxysporum* f. sp. *lactucae*. Lettuce blight is a typical soil-transmitted disease that affects the roots of lettuce, causing the plant to wilt and, in severe cases, to die. The rapid spread and prevalence of lettuce blight and its occurrence in large areas are caused by the lack of disease-resistant varieties, flood irrigation of low borders, improper and uncentralized cultivation of seedlings, improper agricultural operation, continuous cropping cultivation, etc. Shooter 101 and Imperial are the main varieties in the production of head lettuce in the northern area and show susceptibility to blight. Therefore, for lettuce breeders, it is an urgent task to introduce, screen, and cultivate new lettuce varieties that are resistant (tolerant) to blight.

5. Root-knot Nematode Disease (*Meloidogyne incognita* Chitwood)

Root-knot nematode disease is a serious disease in lettuce production. Generally, it results in yield drop of 10%–20%. In severe cases, it results in yield drop of higher than 75% or even no yield at all. This disease can occur in both open and protected plots in the South and mostly in protected plots in the North. At the onset of the disease, symptoms of aboveground parts are not obvious, and outer leaves are withered only at noon when it is hot and dry. In severe cases, the aboveground parts of seedlings or plants are short and grow slowly and weakly; the head lettuce does not form a head or the head is loose; the leaf color is light or even yellow, milky beaded or sugar-gourd enlarged nodules can be observed at the roots, and the root knots darken and gradually decay at a late stage. Root-knot nematodes overwinter in the soil in the form of second-instar larvae or eggs and become the initial source of infection when conditions are appropriate. They invade from the young root tips of lettuce, stimulate cell division, and expand to form nodular root knots; the larvae develop into imagoes inside the root knots, and mate to lay eggs or propagate through parthenogenesis to form a new disease cycle. There are four main species of root-knot nematodes known to harm vegetables: *M. incognita, M. javanica, M. arenaria,* and *M. hapla*, with *M. incognita* being the most harmful. Due to the limited germplasm resources of root-knot nematode-resistant plants, there are nematode-resistant varieties for only a few vegetable crops, including tomato, pepper, and cowpea. Little research has been published on lettuce varieties resistant to root-knot nematodes, and there is no resistant variety that can be applied in practical production in China. With the increasing market demand for lettuce at home and abroad and the progress of molecular biology techniques, the breeding of lettuce resistant (tolerant) to root-knot nematodes will become a trend and focal point.

6. Leaf Blight (Tipburn)

Leaf blight is a physiological disease that tends to occur in both open fields and greenhouses. There are at least five conditions conducive to the disease: high temperature and high humidity, excessive nutrition, excessive water, sudden change of temperature from high to low before harvest, and uneven nutrients, especially calcium deficiency. The main symptom of leaf blight is the appearance of dark brown spots at the margins of the wrapping leaves, which later develop into expanded and interconnected brown spots, resulting in a necrotic outer margin of the entire leaf. The Great Lake varieties have stronger resistance to leaf blight than the Imperial ones. The Great Lake 659 has the strongest resistance in particular. Leeper successfully developed Val Prize, a crisp-leaf head variety with strong resistance to leaf blight in 1991.

V. Breeding for Heat and Bolting Resistance

Leaf lettuce is native to the Mediterranean coast and prefers a cold climate with a proper growth temperature of 15–20°C. They cannot grow well beyond 30°C with decreased edible. Moreover, the heat tolerance of leaf lettuce is significantly correlated with bolting: Varieties with low heat resistance are highly susceptible to bolting. Therefore, the breeding of heat-resistant varieties is equivalent to the breeding of bolting-resistant varieties. In most areas of China, the temperature is often above 30°C in the open field in summer and higher in the protected field. High-temperature stress becomes the main factor hindering the summer cultivation and annual supply of leaf lettuce. Therefore, it is more important to select and cultivate varieties with relatively strong heat resistance. It has been shown that the leaf shape index and stem internode length of functional leaves at the seedling stage are closely related to heat resistance.

VI. Breeding of Varieties for Greenhouses

The breeding of varieties for greenhouses is mainly carried out in Europe, and the varieties developed are mostly of the soft-leaf head (Butterhead) type. Given the low latitudes, long sunshine duration in winter, high rainfall, and low temperature, countries such as the Netherlands, Britain, and France are suitable for greenhouse cultivation of leaf lettuce. At present, such varieties including Wintersalad, Greenway, Valmaihe, Parris, and Island have been successively developed. These varieties are all suitable for greenhouse production with high yield, good quality, and resistance to cold and downy mildew.

Task 3 Selective Breeding and Sexual Crossbreeding

Since lettuce is a self-pollinated crop, hybridization is difficult and costly. Even if lettuce has certain heterosis, it is difficult to promote large-scale hybrid seed production. Therefore, classical selective breeding and conventional sexual crossbreeding are adopted, that is, selecting the superior lettuce plants from the natural population with rich variation, or crossbreeding different types of lettuce, then self-crossing, and selecting according to the traits of subsequent self-crossed generations.

I. Selective Breeding

Although lettuce is a self-pollinated crop, the outcrossing rate is also about 1%, and the natural hybridization rate will increase under a dry climate. Therefore, field variants generated through occasional hybridization or gene mutation can be used to select superior individual plants to obtain genetically stable new varieties. The selection process can be simply described as follows: Varieties and lines with a high variation rate and many superior variations are used as the source of superior individual plants for self-crossing. In the next generation, each strain is planted in one plot, and strains are compared. Superior individual plants are then selected from the superior strains for self-crossing. Qualitative traits are selected in the early stage and quantitative traits in the late stage. The selection continues for many generations until a variety appears with little difference between individual plants and stable traits in the upper and lower generations. Once such a variety is obtained, the variety comparison test, region test, and production test can be carried out. The specific selection criteria should be based on breeding objectives, and the selection should be carried out according to traits such as lettuce quality, high yield, disease and stress resistance, and bolting resistance.

Most of the current lettuce varieties in China are local varieties or new varieties developed by selective breeding from local varieties. For example, the Feiqiao lettuce No. 1, developed by the seed banks of Yong'an and Sanming, Fujian Province, is selected and bred from the local variety Feiqiao lettuce. Compared with the original local variety, the yield of this variety has increased by more than 20%, and the harvest time is 5 days earlier.

II. Sexual Crossbreeding

Another way of lettuce breeding is through the purposeful selection of parents according to breeding objectives and trait inheritance rules to create variation through sexual crossbreeding. Pedigree breeding is the main method used in sexual crossbreeding. Starting from F_2, strain selection is carried out among large populations, and later on, strain selection continues among the superior lines. Generally, superior and stable lines with good traits can be obtained at F_5. As the in-line segregation since F_2 decreases with each generation of self-crossing, the size of the population at pedigree selection is reduced by 50% per generation. As the number of generations increases, the selection focus should shift from the performance of individual plants in early generations to systematic performance in higher generations. In practice, selection in early generations should focus on traits with high heritability, while in high generations, the focus of selection can shift to traits with low heritability. Therefore, a broad hereditary basis should be maintained in early generations so that systems of late generations maintain a sufficient variation rate.

In the process of lettuce crossbreeding, backcrossing and self-crossing are usually combined to enhance the desired superior traits and eliminate undesirable traits. When the backcross method is applied, the backcross parent should only have the traits with a simple mode of inheritance to be modified, while the traits that need to be transferred can maintain a certain expression level in the segregating population. With the increase of backcross times, the genotypes of backcross offspring continue to return to the original state (recurrent parents). Therefore, plants that are similar to backcross parents yet with their own unique traits that are just needed for breeding should be selected for backcross where possible.

Since lettuce is a self-pollinated vegetable with a high planting density, bulk-population selection should be used for offspring selection in sexual hybridization. Most of the new types of leaf lettuce in the United States, Japan, and Europe are developed through conventional sexual hybridization (including distant hybridization).

Task 4 Breeding of Superior Varieties

I. Breeding Methods and Techniques of Conventional Varieties

1. Production of Stock Seeds

Like other vegetables, the selection of superior individual plants and purification can be adopted for the production of lettuce stock seeds. In the stock seed field or superior seed harvesting field, select superior individual plants and cover them with yarn nets or paper bags to prevent outcrossing. After the seeds are mature, collect them from each individual plant separately. Respectively sow the seeds in the year harvested or the next year, observe and compare the plants, select superior strains, and then select superior individual plants among them. Superior strains can be selected after continuous individual plant selection of 2–3 generations. Then, a mixed seed collection will be performed to obtain stock seeds. In breeding stock seeds, strict requirements should be implemented for selection. Good cultivation management measures should be taken to ensure superior stock seeds are available for the breeding of production seeds.

2. Planting of Seed Plants and Field Management

Lettuce is a self-pollinated crop, but natural hybridization can also occur. Different variants and varieties should be spaced more than 500 m apart during planting. The sowing and seedling stage management of lettuce seed plants are basically the same as those of vegetable cultivation. To ensure a high seeding rate, the sowing amount can be appropriately increased, and the seedlings can be planted with a row space of (40–50) cm × (30–40) cm when they have 5–6 true leaves. To improve the survival rate, the main root, which is about 6 cm in length, should be kept upon seedling digging. If the main root is left too short, there will be few lateral roots after planting, which is not easy for seedling recovery.

After the planted seed plants survive, apply nitrogen fertilizer for the first time in combination with watering, and then perform intertillage and hardening of seedlings to promote the formation of a strong root system and a lush leaf cluster. When the stem starts to expand, water it promptly and apply nitrogen fertilizers for the second time. During the vigorous growth period of the stems and leaves, apply nitrogen and potash fertilizers for the third time to facilitate the growth in the late period. Before the seed plants show buds, appropriately control the fertilizer and water to reduce cracked heads, cracked stems, and rot. During the flowering period of seed plants, ensure sufficient fertilizer and water. At the end of the flowering period, erect stands and reduce watering to promote seed maturation. In addition, to prevent rot, some of the old leaves from the lower part of the seed plants should be removed to facilitate ventilation.

3. Seed Plant Selection and Seed Collection

To ensure variety purity and improve seed quality, multiple selections and eliminations should be performed during lettuce seed collection. Weak, diseased, and abnormal plants should be eliminated during seedling thinning, while robust high-quality seedlings that meet the characteristics of the variety should be selected for planting. When the vegetative growth ends, inferior and abnormal plants can be eliminated during the picking of commercial vegetables, and typical, disease-free, thick-stem and late-bolting plants should be kept for seed collection.

It is often hot and rainy around the time of the maturation of lettuce seeds. During this period, the seeds are easy to fly away due to their umbrella-shaped fine hair. In addition, seeds on different parts of the plants vary greatly in terms of the maturity period. Therefore, it is best to collect the seeds in batches so as not to affect the seed yield and quality. If it is difficult to do so, considering the large area of plants cultivated, one-time collection can be performed when the leaves of the seed plants turn yellow and a white umbrella-shaped pappus grows on the seeds. The seeds should be collected after going through an after-ripening process.

II. Seed Storage and Processing

1. Seed Testing

Seed testing, also known as seed identification, is to determine the quality and use value of crop seeds and promote agricultural production through the field and indoor analysis and identification, and inspection of the variety purity, seed purity, germination rate, and water content, as well as disease harm and weed seeds.

Lettuce seed testing involves testing of both variety quality and sowing quality. The variety quality includes both authenticity and purity of the variety, while the sowing quality refers to the germination rate, seed purity, dry grain weight, and water content of the seeds.

Lettuce seeds are classified into stock seeds and field seeds. The former represents a variety purity not lower than 99.0% and the latter not lower than 95.0%. They are the same in terms of other indexes: The purity is not less than 96.0%, the germination rate is not lower than 80%, and the water content is not higher than 7.0%.

2. Seed Treatment

After harvesting, lettuce seeds are generally treated for easy storage and preservation. Such treatment includes cleaning, drying, disinfection, and packaging. Seed cleaning is to remove straw, broken leaves, pericarp, gravels, sediment, sclerotia, and weed seeds mixed in the seed population. It can be carried out by blowing, sieving, or a combination of both. Seed drying helps avoid seed heating, deterioration, and rot during storage. It is realized by spreading the seeds out to let them dry naturally, heating them mechanically, or using desiccants. Lettuce seeds are small and easy to dehydrate. At present, the ultra-dry storage method is used to reduce the water content of seeds to below 5%, which has a good effect on lettuce seeds. It has been found that the germination rate of lettuce seeds with a water content of 3.13% and 2.71% can be maintained at around 92% after 1 year of storage. Seed disinfection can be carried out using some chemicals such as captan and thiram. To prevent the mixing of varieties, ensure safe storage and transportation, and ease of sale, seeds should be packaged. At present, common packages used include jute bags, cloth bags, tin cans, small paper bags; aluminum foil composite bags, and polyethylene bags. Jute bags and cloth bags are suitable for short-term storage of a large number of seeds; tin cans are suitable for long-term storage of a small number of seeds or market sales; small paper bags, polyethylene bags, aluminum foil composite bags, and other bags are suitable for packaging a small number of retail seeds. It should be noted that there is a certain drying requirement for lettuce seeds to be sealed in airtight containers: The upper limit of water content must be 5.5%.

3. Seed Storage

Seed storage helps to maintain the high viability of seeds for a long time, prolong the service life of seeds, and ensure the high variety and sowing quality of seeds, so as to meet the requirements of production for seed quantity and quality. Lettuce seeds are not resistant to storage, and the germination rate drops rapidly after they are stored in common warehouses for half a year. Studies show that lettuce seeds can be packed in cloth bags and stored in cold storage if conditions permit. This saves costs and reduces the water content of seeds. If seeds are to be stored at room temperature, they need to be dried until they have medium or low water content, and packaged in aluminum bags to maintain a constant low water content. When the cold storage capacity is limited or in order to reduce the use of cold storage to lower costs, the cold storage can be used for temporary storage: Pack the lettuce seeds in cloth bags and store them in cold storage, dry the seeds in batches when it is sunny, and then pack them in aluminum bags and store them at room temperature.

[Summary]

I. Key Points

Breeding techniques and methods of conventional varieties, seed testing methods, and storage methods.

II. Difficult Points

Master the storage and processing methods of lettuce seeds.

[Skill Training]

Skill Training 8-1: Lettuce Seed Purity Analysis

I. Purposes and Requirements

(1) Master seed purity analysis techniques;

(2) Be able to correctly identify pure seeds, other crop seeds, and inert matter.

II. Materials and Tools

Test sample, purity analysis workbench, sample divider, sample dividing plate, set screen, platform scale with a sensitivity of 0.1, balances with a sensitivity of 0.01 and 0.001, respectively, small discs or small plates, tweezers, small scrapers, magnifying glass, small brushes, blowers, etc.

III. Test Procedure

1. Inspection of Heavy Mixed Materials

Pick out the heavy mixed materials from the test sample, weigh and classify them into heavy mixed materials of other crop seeds and inert matter, and weigh them respectively.

2. Test Sample Dividing

Take one portion or half of the test sample with a sample divider, and weigh them.

3. Analysis, Separation, and Weighing of Test Sample

Pour separated materials on the purity analysis workbench respectively through sieving, and divide the test samples into pure seeds, other crop seeds, and inert matter manually or with the help of certain instruments (such as magnifying glass, binocular dissecting microscopes, and seed blowers), and respectively put them into the corresponding containers and weigh them.

4. Result Calculation

Calculate the weight gain/loss percentage and the weight percentage of each component, and check the allowable difference and percentage round-off.

5. Determination of the Number of Other Crop Seeds

Take a specified number of all the remaining portions or the other half of the test sample; spread them on the purity analysis workbench or sample disc. Observe seed by seed to find out all other crop seeds or designated seeds, count the number of seeds of each variety, and then add them to the corresponding number of seeds in the previous portion or first half of the test sample. The result is expressed in the number of seeds per unit test sample weight.

IV. Skill Training Report

Fill out the report (Table 8-1) on the purity analysis results and write down the calculation process.

Table 8-1 Report on Purity Analysis Results Sample No.

Crop Name		Scientific Name	
Component	Pure Seeds	Other Crop Seeds	Inert Matter
Percentage			
Name and number of other crop seeds or content per kg (specify scientific name)			
Remarks			

Tested by: Date:

Skill Training 8-2: Visiting Plant Tissue Culture Laboratory

I. Purposes and Requirements

(1) Be familiar with the plant tissue culture process;

(2) Master the basic facilities and design requirements for plant tissue culture laboratory;

(3) Be familiar with the instruments, equipment, and tools commonly used for plant tissue culture.

II. Experimental Equipment and Facilities

1. Basic Facilities

Plant tissue culture laboratory or tissue culture factory.

2. Instruments and Equipment

Clean benches, autoclaves, distilled water generators or pure water generators, filling machines, acidimeters, bottle washers, common refrigerators, water heaters, microwave ovens, electric furnaces, dissecting microscopes, balances, incubators, combustion boxes, centrifuges, shakers, light culture boxes, light culture racks, air conditioners, and other equipment, various culture dishes (including glass Petri dishes), and tools.

III. Test Procedure

(1) The tutor introduces the purposes, requirements, and tasks of the experiment training.

(2) The tutor introduces rules and related precautions for operation in the plant tissue culture laboratory, or the management of the tissue culture factory introduces relevant rules and regulations of the factory.

(3) Divide the whole class into several groups according to the class size to be led and guided by the tutor and the trainer, respectively.

(4) Visit the laboratory according to the plant tissue culture process. Introduce the room layout of the laboratory, basic facilities, functions, and design requirements of each room.

(5) Introduce to each group the names of Petri dishes, instruments, and equipment in each room and their purposes.

(6) Visit the greenhouse where test-tube seedlings are to be acclimatized and transplanted, and observe the growth of such seedlings transplanted.

IV. Training Report

(1) Draw the room layout of a plant tissue culture laboratory (or tissue culture factory).

(2) List the names of instruments, equipment, and dishes commonly used in a plant tissue culture laboratory (or tissue culture factory) and their purposes.

Skill Training 8-3: Factor Analysis for Introduction of Horticultural Plants

I. Purposes and Requirements

(1) Enable trainees to have a deeper understanding of the theoretical knowledge about the introduction of horticultural plants through analysis of basic influencing factors;

(2) Enable trainees to get familiar with the main aspects of introduction;

(3) Improve trainees' practical ability in organizing and leading introduction work.

II. Research Content

Comprehensively collect and analyze relevant information about the place of origin of materials and the place to which they will be introduced, compare the various factors affecting the introduction results, and comprehensively demonstrate the important factors affecting the introduction. Take into full account the climatic conditions of the place where the plants will be introduced, and the areas from which the plants will be chosen to be introduced, with similar latitudes, altitudes, and soil conditions. In addition, the adaptability of the plants to be introduced, the planting management of the place where the plants are to be introduced, human factors, and other factors should be considered. The adaptability of horticultural plants is related to the ecological environment of the current area and the ecological conditions through their phylogeny.

III. Test Procedure

1. Collect and Sort Data

Before the start of training, arrange for trainees to collect or research the following data, and sort them:

(1) Data on the distribution of the material to be introduced, the significance for cash crop cultivation, the biological characteristics and phylogeny, among others;

(2) Data on the geography, climate, soil, and vegetation of the place of origin of the material to be introduced;

(3) The experience and summary reports of the successful introduction of the material.

2. Analyze and Compare the Main Factors Affecting the Introduction

According to various data on the material to be introduced, analyze and compare the similarity of various factors concerning the place of origin and the place for the plants to be introduced, and find out the restraints affecting the successful introduction. For example, latitude, elevation, climate (light, temperature, humidity, and rainfall), soil, vegetation, as well as cultivation history, cultivation management, and economic development level.

3. Exchange and Demonstrate

(1) Demonstrate the necessity of the introduction. Explain the necessity for the introduction of horticultural plants by combining the theoretical knowledge of the introduction learned in class and the relevant data collected.

(2) Demonstrate the feasibility of the introduction. Based on the biological

characteristics, place of origin, or natural distribution area of the material to be introduced, compare and analyze the geographical and ecological factors in the place to which the material will be introduced. Exchange ideas with each other and demonstrate the feasibility of the introduction.

(3) Demonstrate the production management measures. According to the economic development and cultivation management of the place to which the material is to be introduced, analyze and demonstrate the cultivation management requirements of the material to be introduced, and propose the corresponding technical measures for cultivation management of the material to be introduced.

IV. Experiment Report

Analyze the basic factors affecting the introduction of horticultural plants.

[Extension Tasks]

I. Review Exercises

(1) What are the main breeding objectives for lettuce? Why?

(2) What should be paid attention to when introducing lettuce?

II. Case Sharing

Application of Biotechnology in Lettuce Breeding

The resistance gene of wild species can be transferred into varieties by using the protoplast fusion technique to overcome the obstacles in sexual hybridization. In the 1990s, the Vegetable and Ornamental Crops Experiment Station in Nagano Prefecture, Japan, fused the cells of varieties and wild species of lettuce and developed the soft-rot-resistant lettuce. Stable plant tissue, cell culture, and regeneration system are important prerequisites for transgenic studies of lettuce through *Agrobacterium tumefaciens*-mediated transformation. Michelmore reported a transgenic system for leaf lettuce in 1987 with a transformation rate of about 10%. In recent years, much genetic improvement has been made on leaf lettuce through Agrobacterium tumefaciens mediation. Singaporean researchers have successfully transplanted a red grape gene that generates resveratrol into red-leaf lettuce. The content of resveratrol is high in red wine, which can effectively reduce harmful cholesterol content, increase beneficial cholesterol content, and play a role in preventing cancer. By injecting the thrombolytic gene of Bacillus subtilis into leaf lettuce, the research team of Gyeongsang National University in Korea, in collaboration with that of Jinju National University of Industry, has developed a leaf lettuce variety that can be used to treat diabetes and stroke. A research institute in Japan has successfully developed transgenic leaf lettuce that can prevent anemia by implanting the ferritin gene in soybeans into leaf lettuce cells. Leaf lettuce has a high content of vitamin C, which facilitates the absorption of iron.